Agricultural and Environmental Sustainability: Considerations for the Future

Agricultural and Environmental Sustainability: Considerations for the Future has been co-published simultaneously as *Journal of Crop Improvement*, Volume 19, Numbers 1/2 (#37/38) 2007.

Monographic Separates from the *Journal of Crop Improvement*®

For additional information on these and other Haworth Press titles, including descriptions, tables of contents, reviews, and prices, use the QuickSearch catalog at http://www.HaworthPress.com.

The *Journal of Crop Improvement*® is the successor title to the *Journal of Crop Production**, which changed title after Volume 9, Numbers 1/2 (#17/18) 2003. The journal under its new title begins as the "*Journal of Crop Improvement*®," Volume 10, Numbers 1/2 (#19/20) 2004.

Agricultural and Environmental Sustainability: Considerations for the Future, edited by Manjit S. Kang, PhD (Vol. 19, No. 1/2 #37/38, 2007). *Examination of the latest studies on the challenges for agricultural and environmental sustainability and food supply security world wide.*

Plant Biotechnology in Ornamental Horticulture, edited by Yi Li, PhD and Yan Pei, PhD (Vol. 17, No. 1/2 #33/34 and Vol. 18, No. 1/2 #35/36, 2006). *A comprehensive overview of the key scientific and technical advances, issues, and challenges in one of the fastest growing segments of the agriculture industry.*

Enhancing the Efficiency of Nitrogen Utilization in Plants, Sham S. Goyal, PhD, Rudolf Tischner, PhD, Amarjit S. Basra, PhD (Vol. 15, No. 2 #30, 2005). *Examines current research on inorganic nitrogen uptake and metabolism in plant life and crop production.*

Genetic and Production Innovations in Field Crop Technology: New Developments in Theory and Practice, Manjit S. Kang, PhD (Vol. 14, No. 1/2 #27/28, 2005). *A comprehensive examination of current research on field crop technology improvements.*

Ecological Responses and Adaptations of Crops to Rising Atmospheric Carbon Dioxide, Zoltán Tuba, DSc (Vol. 13, No. 1/2 #25/26, 2005). *Examines in detail the ecophysiological responses of crops to elevated air carbon dioxide and the economic significance of these changes.*

New Dimensions in Agroecology, David Clements, PhD and Anil Shrestha, PhD (Vol. 11, No. 1/2 #21/22 and Vol. 12, No. 1/2 #23/24, 2004). *Provides extensive information on current innovative agroecological research and education as well as emerging issues in the field.*

Adaptations and Responses of Woody Plants to Environmental Stresses, edited by Rajeev Arora, PhD (Vol. 10, No. 1/2 #19/20, 2004). *Focuses on low-temperature stress biology of woody plants that are of horticultural importance.*

Cropping Systems: Trends and Advances, edited by Anil Shrestha, PhD* (Vol. 8, No. 1/2 #15/16 and Vol. 9, No. 1/2 #17/18, 2003). *"Useful for all agricultural scientists and especially crop and soil scientists. Students, professors, researchers, and administrators will all benefit. . . . THE CHAPTER AUTHORS INCLUDE PRESENT AND FUTURE LEADERS IN THE FIELD with broad international perspectives. I am planning to use this book in my own teaching" (Gary W. Fick, PhD, Professor of Agronomy, Cornell University)*

Crop Production in Saline Environments: Global and Integrative Perspectives, edited by Sham S. Goyal, PhD, Surinder K. Sharma, PhD, and D. William Rains, PhD* (Vol. 7, No. 1/2 #13/14, 2003). *"TIMELY. . . . COMPREHENSIVE. . . . The authors have considerable experience in this field. I hope this book will be read widely and used for promoting soil health and sustainable advances in crop production." (M. S. Swaminathan, PhD, UNESCO Chair in Ecotechnology, M. S. Swaminathan Research Foundation, Chennai, Tami Nadu, India)*

Food Systems for Improved Human Nutrition: Linking Agriculture, Nutrition, and Productivity, edited by Palit K. Kataki, PhD and Suresh Chandra Babu, PhD* (Vol. 6, No. 1/2 #11/12, 2002). *Discusses the concepts and analyzes the results of food based approaches designed to reduce malnutrition and to improve human nutrition.*

Quality Improvement in Field Crops, edited by A. S. Basra, PhD, and L. S. Randhawa, PhD* (Vol. 5, No. 1/2 #9/10, 2002). *Examines ways to increase nutritional quality as well as volume in field crops.*

Allelopathy in Agroecosystems, edited by Ravinder K. Kohli, PhD, Harminder Pal Singh, PhD, and Daizy R. Batish, PhD* (Vol. 4, No. 2 #8, 2001). *Explains how the natural biochemical interactions among plants and microbes can be used as an environmentally safe method of weed and pest management.*

The Rice-Wheat Cropping System of South Asia: Efficient Production Management, edited by Palit K. Kataki, PhD* (Vol. 4, No. 1 #7, 2001). *This book critically analyzes and discusses production issues for the rice-wheat cropping system of South Asia, focusing on the questions of soil depletion, pest control, and irrigation. It compiles information gathered from research institutions, government organizations, and farmer surveys to analyze the condition of this regional system, suggest policy changes, and predict directions for future growth.*

The Rice-Wheat Cropping System of South Asia: Trends, Constraints, Productivity and Policy, edited by Palit K. Kataki, PhD* (Vol. 3, No. 2 #6, 2001). *This book critically analyzes and discusses available options for all aspects of the rice-wheat cropping system of South Asia, addressing the question, "Are the sustainability and productivity of this system in a state of decline/stagnation?" This volume compiles information gathered from research institutions, government organizations, and farmer surveys to analyze the impact of this regional system.*

Nature Farming and Microbial Applications, edited by Hui-lian Xu, PhD, James F. Parr, PhD, and Hiroshi Umemura, PhD* (Vol. 3, No. 1 #5, 2000). *"Of great interest to agriculture specialists, plant physiologists, microbiologists, and entomologists as well as soil scientists and evnironmentalists. . . . very original and innovative data on organic farming." (Dr. André Gosselin, Professor, Department of Phytology, Center for Research in Horticulture, Université Laval, Quebec, Canada)*

Water Use in Crop Production, edited by M.B. Kirkham, BA, MS, PhD* (Vol. 2, No. 2 #4, 1999). *Provides scientists and graduate students with an understanding of the advancements in the understanding of water use in crop production around the world. You will discover that by utilizing good management, such as avoiding excessive deep percolation or reducing runoff by increased infiltration, that even under dryland or irrigated conditions you can achieve improved use of water for greater crop production. Through this informative book, you will discover how to make the most efficient use of water for crops to help feed the earth's expanding population.*

Expanding the Context of Weed Management, edited by Douglas D. Buhler, PhD* (Vol. 2, No. 1 #3, 1999). *Presents innovative approaches to weeds and weed management.*

Nutrient Use in Crop Production, edited by Zdenko Rengel, PhD* (Vol. 1, No. 2 #2, 1998). *"Raises immensely important issues and makes sensible suggestions about where research and agricultural extension work needs to be focused." (Professor David Clarkson, Department of Agricultural Sciences, AFRC Institute Arable Crops Research, University of Bristol, United Kingdom)*

Crop Sciences: Recent Advances, Amarjit S. Basra, PhD* (Vol. 1, No. 1 #1, 1997). *Presents relevant research findings and practical guidance to help improve crop yield and stability, product quality, and environmental sustainability.*

Agricultural and Environmental Sustainability: Considerations for the Future

Manjit S. Kang, PhD
Editor

Agricultural and Environmental Sustainability: Considerations for the Future has been co-published simultaneously as *Journal of Crop Improvement*, Volume 19, Numbers 1/2 (#37/38) 2007.

Haworth Food & Agricultural Products Press™
An Imprint of The Haworth Press, Inc.

New York • London • Victoria (AU)
www.HaworthPress.com

Published by

Haworth Food & Agricultural Products Press™, 10 Alice Street, Binghamton, NY 13904-1580
USA

Haworth Food & Agricultural Products Press™ is an imprint of The Haworth Press, Inc., 10 Alice Street, Binghamton, NY 13904-1580 USA.

Agricultural and Environmental Sustainability: Considerations for the Future has been co-published simultaneously as *Journal of Crop Improvement*, Volume 19, Numbers 1/2 (#37/38) 2007.

Library of Congress Cataloging-in-Publication Data

Agricultural and environmental sustainability : considerations for the future / Manjit S. Kang, editor.
   p. cm.
   "Agricultural and environmental sustainability: considerations for the future has been co-published simultaneously as Journal of crop improvement, volume 19, numbers ½ (#37/38) 2007."
   Includes bibliographical references and index.
   ISBN-13: 978-1-56022-170-8 (hard cover : alk. paper)
   ISBN-10: 1-56022-170-4 (hard cover : alk. paper)
   ISBN-13: 978-1-56022-171-5 (soft cover : alk. paper)
   ISBN-10: 1-56022-171-2 (soft cover : alk. paper)
   1. Sustainable agriculture. 2. Agriculture–Environmental aspects. I. Kang, Manjit, S.
S494.5.S86A366 2007
630–dc22
                       2006034074

The HAWORTH PRESS Inc.
Abstracting, Indexing & Outward Linking
PRINT and ELECTRONIC BOOKS & JOURNALS

This section provides you with a list of major indexing & abstracting services and other tools for bibliographic access. That is to say, each service began covering this periodical during the the year noted in the right column. Most Websites which are listed below have indicated that they will either post, disseminate, compile, archive, cite or alert their own Website users with research-based content from this work. (This list is as current as the copyright date of this publication.)

Abstracting, Website/Indexing Coverage . . . . Year When Coverage Began

- *(CAB ABSTRACTS, CABI) <http://www.cabi.org>* **2006**
- ***Academic Search Premier (EBSCO)****
  <http://www.epnet.com/academic/acasearchprem.asp> **2006**
- ***Chemical Abstracts Service** <http://www.cas.rg>* **1998**
- ***MasterFILE Premier (EBSCO)****
  <http://www.epnet.com/government/mfpremier.asp> **2006**
- *ABAFR (Aquatic Biology, Aquaculture and Fisheries Resources) (NISC USA) <http://www.nisc.com>* **2006**
- *Academic Source Premier (EBSCO)* **2007**
- *AFDEV (African Development Database) (NISC SA) <http://www.nisc.co.za>* **2006**
- *AfricaWide-NiPAD (NISC SA) <http://www.nisc.co.za>* **2006**
- *AgBiotech News & Information (CAB ABSTRACTS, CABI) <http://www.cabi.org>* **2006**
- *AGRICOLA Database (National Agricultural Library) (AGRICultural OnLine Access) A Bibliographic database of citations to the agricultural literature created by the National Agricultural Library and its cooperators. <http://www.natl.usda.gov/ag98>* **2006**
- *Agricultural Engineering Abstracts (CAB ABSTRACTS, CABI) <http://www.cabi.org>* **2006**

(continued)

(continued)

(continued)

(continued)

Bibliographic Access

- *MediaFinder <http://www.mediafinder.com>*

- *Ulrich's Periodicals Directory: The Global Source for Periodicals Information Since 1932 <http://www.bowkerlink.com>*

Special Bibliographic Notes related to special journal issues (separates) and indexing/abstracting:

- indexing/abstracting services in this list will also cover material in any "separate" that is co-published simultaneously with Haworth's special thematic journal issue or DocuSerial. Indexing/abstracting usually covers material at the article/chapter level.
- monographic co-editions are intended for either non-subscribers or libraries which intend to purchase a second copy for their circulating collections.
- monographic co-editions are reported to all jobbers/wholesalers/approval plans. The source journal is listed as the "series" to assist the prevention of duplicate purchasing in the same manner utilized for books-in-series.
- to facilitate user/access services all indexing/abstracting services are encouraged to utilize the co-indexing entry note indicated at the bottom of the first page of each article/chapter/contribution.
- this is intended to assist a library user of any reference tool (whether print, electronic, online, or CD-ROM) to locate the monographic version if the library has purchased this version but not a subscription to the source journal.
- individual articles/chapters in any Haworth publication are also available through the Haworth Document Delivery Service (HDDS).

As part of Haworth's continuing committment to better serve our library patrons, we are proud to be working with the following electronic services:

AGGREGATOR SERVICES

EBSCOhost

Ingenta

J-Gate

Minerva

OCLC FirstSearch

Oxmill

SwetsWise

Ingenta
MINERVA
FirstSearch
Oxmill Publishing
SwetsWise

LINK RESOLVER SERVICES

1Cate (Openly Informatics)

CrossRef

Gold Rush (Coalliance)

LinkOut (PubMed)

LINKplus (Atypon)

LinkSolver (Ovid)

LinkSource with A-to-Z (EBSCO)

Resource Linker (Ulrich)

SerialsSolutions (ProQuest)

SFX (Ex Libris)

Sirsi Resolver (SirsiDynix)

Tour (TDnet)

Vlink (Extensity, *formerly Geac*)

WebBridge (Innovative Interfaces)

LinkOut.

SerialsSolutions

SirsiDynix
TOUR
extensity
WebBridge

Agricultural and Environmental Sustainability: Considerations for the Future

CONTENTS

ABOUT THE EDITOR

Manjit S. Kang, PhD, Professor of Quantitative Genetics at the Louisiana State University Agricultural Center and LSU A&M campuses in Baton Rouge, received his PhD degree in 1977 in Crop Science (Genetics and Plant Breeding) from University of Missouri-Columbia. He has edited/authored 10 books, including *Genotype-by-Environment Interaction, Quantitative Genetics, Genomics and Plant Breeding, Crop Improvement: Challenges in the Twenty-First Century, Handbook of Formulas for Plant Geneticists and Breeder, GGE Biplot Analysis*, and *Genetic and Production Innovations in Field Crop Technology.*

Dr. Kang lectured on quantitative genetics as applied to crop improvement in Hungary under the USDA/OICD sponsorship in 1992. In 1999, he was the recipient of a U.S. Department of State's prestigious Fulbright Senior Scholar Award to lecture on quantitative genetics and statistical genomics at Malaysian universities (June-August). He was the recipient of the 2005 South Dakota Crop Improvement Association Lectureship Award. He has conducted workshops/seminars at International Rice Research Institute in the Philippines; International Institute of Tropical Agriculture in Nigeria; and Yunnan Academy of Agricultural Sciences, Kunming, China. He has regularly taught a course on "Quantitative Genetics in Plant Improvement" at Louisiana State University since 1986. In addition, he has taught courses on "GGE Biplot Analysis," "Advanced Plant Genetics," and "Advanced Plant Breeding" at LSU. Previously, he taught Plant Genetics and Plant Breeding courses at Southern Illinois University-Carbondale in the 1970s.

Dr. Kang has organized successful international symposia on genotype-by-environment interaction in plant breeding, quantitative genetics and genomics, and agricultural and environmental sustainability. He has published more than 100 refereed journal articles and many book chapters and essays in encyclopedias of genetics and crop science. He has been active in Sigma Xi since 1975 and served as President of the LSU Chapter of Sigma Xi twice (2000-2001 and 2005-2006). He was recently selected as a Sigma Xi Distinguished Lecturer (2007-2009). He served as President of the Association of Agricultural Scientists of Indian Origin from 2003 to 2006.

Dr. Kang is Fellow of American Society of Agronomy and Crop Science Society of America. He serves as Technical Editor of *Crop Science*, Editor of *Journal of Crop Improvement*, and Editor-in-Chief of *Communications in Biometry and Crop Science*. The Punjab Agricultural University-Ludhiana, India, recognized him for his significant contributions to plant breeding and genetics at its 36th Foundation Day in 1997. Dr. Kang's biographical sketches have appeared in Marquis Who's Who.

Foreword

Until the 1972 Stockholm Conference on the Human Environment, sustainability was primarily measured in economic terms. The Stockholm Conference and later the Brundtland Report on "Our Common Future," as well as the United Nations Conference on Environment and Development held at Rio de Janeiro in 1992 led to the introduction of environmental criteria in the assessment of sustainability. The sciences of ecological economics and economic ecology were thus born. The U.S. National Academy of Sciences has also introduced a section on "Sustainability Science" in its Proceedings. The Johannesburg Summit reinforced the need for mainstreaming ecological parameters in the measurement of sustainability.

At Rio de Janeiro, the World Business Council on Sustainable Development emphasized that good ecology is also good business. Conversely, neglect of environmental factors, like the impact of technology and development on land, water, forest, biodiversity and climate, would lead to the collapse of both industry and agriculture in the long term. Ecological economics measures profits and income from a long-term time dimension. The concept of inter-generational equity emphasizes that we should take into account the impact of development projects on the generations yet to be born.

In addition to economics and ecology, it is now clear that the social dimensions of sustainability should also be kept in view. Where hunger rules, peace cannot prevail. This is why the United Nations Millennium Development Goals have placed emphasis on the elimination of hunger and poverty and the promotion of gender equity and environmental safety.

[Haworth co-indexing entry note]: "Foreword." Swaminathan, M. S. Co-published simultaneously in *Journal of Crop Improvement* (Haworth Food & Agricultural Products Press, an imprint of The Haworth Press, Inc.) Vol. 19, No. 1/2 (#37/38), 2007, pp. xxv-xxvi; and: *Agricultural and Environmental Sustainability: Considerations for the Future* (ed: Manjit S. Kang) Haworth Food & Agricultural Products Press, an imprint of The Haworth Press, Inc., 2007, pp. xvii-xviii. Single or multiple copies of this article are available for a fee from The Haworth Document Delivery Service [1-800-HAWORTH, 9:00 a.m. - 5:00 p.m. (EST). E-mail address: docdelivery@haworthpress.com].

xvii

Agricultural and Environmental Sustainability: Considerations for the Future will have to be assessed from the economic, ecological and social viewpoints. This book is, therefore, a timely and important contribution to the cause of generating awareness of sustainability issues among scientists and scholars. Global warming and sea level rise are no longer in the realm of speculation. Changes in temperature and precipitation as well as the more frequent occurrence of cyclonic storms have now become a reality. I am confident that the articles in this book will show the way to ensuring that development leads to enduring happiness in human societies. We are indebted to Dr. Manjit S. Kang for his labor of love for the cause of sustainable development, which has resulted in the present publication. I hope it will be read widely and its principal messages will be integrated in the design of research and development programs in the area of agriculture and food security. Food security is best defined as "physical, economic, ecological and social access to balanced diet and safe drinking water." Such a definition will help provide impetus to scientists and policy makers for making environmental, economic and social sustainability the bottom line of their work.

M. S. Swaminathan
Chairman, National Commission on Farmers, Government of India
President, Pugwash Conferences on Science and World Affairs
Chairman, M. S. Swaminathan Research Foundation
Third Cross Street, Taramani Institutional Area
Chennai-600 113 (India)
E-mail: swami@mssrf.res.in; msswami@vsnl.net

Preface

There are various definitions and concepts of sustainability. A sustainable system is one that provides for the needs of the world population without jeopardizing the ability of future generations to provide for them. A sustainable system can be carried out again and again without negatively impacting the environment. Thus, sustainable agriculture is a process of producing food that is healthy for consumers and animals, does not harm the environment, is humane for workers and animals, provides a fair wage to the farmer, and supports and enhances rural communities (see http://www.sustainabletable.org/intro/whatis/).

According to the U.S. Department of Agriculture, food security means access by all people at all times to enough food for an active, healthy life. Food security includes at a minimum: (1) ready availability of nutritionally adequate and safe foods, and (2) an assured ability to acquire acceptable foods in socially acceptable ways. The phrase 'food security' is often used to mean a situation in which people have continuity of food supply. According to the United Nations Food and Agriculture Organization (FAO), around 828 million men, women and children are chronically hungry while 2 billion people lack food security because of poverty. On the occasion of the annual observance of *World Food Day* on 16 October 2005, the FAO Director-General Jacques Diouf said in Rome, "Today the world has the resources and technology to produce sufficient quantities of food not only to meet the demand of a growing population, but also to bring an end to hunger and poverty." According to the IITA Director General, Dr. P. Hartmann, one of the most effective ways to alleviate poverty, and in turn its inseparable partner hunger, is through agriculture and the production of more food.

[Haworth co-indexing entry note]: "Preface." Kang, Manjit S. Co-published simultaneously in *Journal of Crop Improvement* (Haworth Food & Agricultural Products Press, an imprint of The Haworth Press, Inc.) Vol. 19, No. 1/2 (#37/38), 2007, pp. xxvii-xxxi; and: *Agricultural and Environmental Sustainability: Considerations for the Future* (ed: Manjit S. Kang) Haworth Food & Agricultural Products Press, an imprint of The Haworth Press, Inc., 2007, pp. xix-xxiii. Single or multiple copies of this article are available for a fee from The Haworth Document Delivery Service [1-800-HAWORTH, 9:00 a.m. - 5:00 p.m. (EST). E-mail address: docdelivery@haworthpress.com].

The authors of this proceedings volume entitled *Agricultural and Environmental Sustainability: Considerations for the Future* are internationally recognized scientists, representing land grant educational institutions and international agricultural centers, such as ICARDA, IRRI, CIMMYT, and IITA. The Foreword is written by Professor M. S. Swaminathan, Chairman, National Commission on Farmers, Government of India; President, Pugwash Conferences on Science and World Affairs; and Chairman, M. S. Swaminathan Research Foundation in Chennai. He points out that neglect of environmental factors, e.g., the impact of technology and development on land, water, forest, biodiversity and climate, would lead in the long-term to the collapse of both industry and agriculture. Dr. Swaminathan advocates inter-generational equity that emphasizes that we should take into account the impact of development projects on the generations yet to be born.

In the first chapter, "Sustainability of Agriculture: Issues, Observations and Outlook," Dr. C. Jerry Nelson discusses various aspects of sustainability of agriculture, such as economic returns, environmental preservation, and sociological factors associated with quality of life. He envisions an expanding definition of sustainability and a new paradigm for agriculture resulting from expectations by the global public regarding the environment, biodiversity, food quality, food safety, and other multifunctional outputs that enhance quality of life.

In his article "Agroforestry for Sustainability of Lower-Input Land-Use Systems," Dr. P. K. Ramachandran Nair points out that agroforestry offers a unique set of opportunities for alleviating poverty and arresting land degradation, and providing ecosystem services in both low-income and industrialized nations. He further points out that improvement and exploitation of the large number of fruit trees and medicinal plants is one of several promising opportunities for enhancing the food and nutritional security without causing the environmental hazards that are characteristic of input-intensive land-use systems.

In the article "Managing Soils for Food Security and Climate Change," Dr. Rattan Lal lists the following issues to be the major ones for developing countries: (i) meeting food demand of the growing population, (ii) reducing risks of soil and ecosystem degradation, (iii) minimizing risks of eutrophication and contamination of natural waters, and (iv) decreasing net emissions of CO_2 and other greenhouse gases. True to his motto "*In soil we trust*," Dr. Lal urges agriculturists not to deplete, degrade and abuse soil but to improve, restore and use it wisely. He suggests that adoption of basic principles of soil management would require a radical shift in the scientific, social, ethnic and cultural fabric of a society.

In their article entitled "Whole-System Integration and Modeling Essential to Agricultural Science and Technology for the 21st Century," Dr. L. R. Ahuja and colleagues elucidate the environmental problems caused by the increased use of fertilizers and pesticides. They suggest that excessive leaching and runoff of agricultural chemicals seriously affect the quality of both the groundwater and surface waters. They give strong arguments for a whole-system modeling approach to greatly enhance the efficiency of field research for developing sustainable agricultural systems, serve as guides for planning and management, and help transfer new technologies applicable to various situations in developing countries.

Dr. S. Rajaram and colleagues discuss in "Sustainability Considerations in Wheat Improvement and Production" global wheat mega-environments and the need for different types of wheat germplasm for those mega-environments. They point out that wheat production and productivity in India have begun to show stagnation because of a decline in natural resource base. They also present evidence of sustainability and profitability of irrigated agriculture in Sonora, Mexico, when farmers practice zero tillage, residue management, and raised-bed planting system.

In "Sustainability of the Rice-Wheat Cropping System: Issues, Constraints, and Remedial Options," Dr. J. K. Ladha and colleagues provide an overview of the rice-wheat cropping system being practiced in the Indo-Gangetic Plains (IGP). They discuss the contribution of this system to food security of the region, the recent significant slowdown in yield growth rate of this system, and the possible causes of the slowdown. They suggest that a paradigm shift is required for enhancing the system's productivity and sustainability.

The next three articles deal with agricultural sustainability issues relative to Africa. Dr. Patrick C. Wall discusses conservation agriculture in "Tailoring Conservation Agriculture to the Needs of Small Farmers in Developing Countries: An Analysis of Issues." He suggests that Conservation Agriculture (CA)–a complex technology–involves not only a change in many of the farmer's cultural practices, but also a change in mind-set to overcome the use of the outmoded plough. His experience indicates that smallholder farmers in Africa generally do not have access to needed links to information systems outside of their local community. Dr. Wall advocates that strategies for the successful adoption and management of CA practices need to address the issue of an enhanced knowledge base of individual farmers and the community.

Drs. B. B. Singh and H. Ajeigbe suggest in their article "Improved Cowpea-Cereals-Based Cropping Systems for Household Food Security and Poverty Reduction in West Africa" that agricultural systems in West Africa are still based on traditional inter-cropping systems with little or no application of fertilizers and chemicals. They reason that this perpetuates malnutrition, hunger and poverty through the vicious circle of 'low input–low production–low income' and food insecurity. They describe a model that involves a holistic combination of new, more productive dual-purpose and resilient cultivars of cowpea, maize, sorghum, and millet in a strip-cropping pattern with a minimum and selective application of fertilizers and pesticides. They present evidence of how model systems have helped farm families become more food secure.

In their article "Bacterial Wilt and Drought Stresses in Banana Production and Their Impact on Economic Welfare in Uganda: Implications for Banana Research in East African Highlands," Drs. S. Abele and M. Pillay discuss implications of stresses caused by disease and drought for banana research and production in Uganda. They investigated the economic impact of *Xanthomonas* wilt (bxw) and drought on banana production, and showed that bxw and drought caused significant losses for both consumers and producers. They suggest that the losses resulting from an extreme scenario seriously jeopardize food security and affect overall macro-economic performance in Uganda. Under more realistic scenarios, i.e., relatively low bxw and drought losses, they still show heavy economic losses but these losses affect mainly consumers. They advocate publicly financed breeding and plant material multiplication and dissemination to benefit farmers.

The final chapter "Greening of Agriculture: Is It All a Greenwash of the Globalized Economy?" is a departure from the ones discussed above in that it mainly deals with developed nations. Dr. Charles Francis and colleagues suggest that there is growing concern about the environmental impacts of agriculture and the food system, and that private companies are eager to exploit this concern by advertising products that are environmentally friendly. They suggest that farms and businesses that measure success as environmental soundness and social responsibility as well as economic returns are likely to be "greener" than ones that only use economics as the single bottom line.

I must mention that this work resulted from a symposium on "*Sustainability of Agriculture, Environment, and Food Security*," which was organized by the Association of Agricultural Scientists of Indian

Origin (AASIO) and held on November 7, 2005 in Salt Lake City, Utah, in conjunction with the annual meetings of the American Society of Agronomy (ASA). The symposium was a celebration of the 25th anniversary of AASIO (1980-2005) and was co-sponsored by ASA Division A6–International Agriculture. The cooperation between the AASIO and Division A6 dates back to 2000-01, when the two organizations joined hands in conducting a symposium on *"Food Security and Sustainable Agricultural Development for the 21st Century in India."* That joint effort resulted in a contribution to the book entitled *"Sustainability of Agricultural Systems in Transition,"* published in 2001 by ASA, CSSA, and SSSA. The scope of the current symposium was broader in the sense that issues discussed related not only to India, but also much of the developing world as well as to developed countries. The editors of *Sustainability of Agricultural Systems in Transition*, W. A. Payne, D. R. Keeney, and S. C. Rao, pointed out that the need to be profitable without depleting non-renewable resources or endangering ecosystems was a common denominator among developing and developed countries.

I would be remiss if I did not express my thanks to all of the symposium speakers and authors of the various articles included in this book. I also want to thank my AASIO colleagues and ASA Division A6 (International Agronomy) Chair, Dr. Andrew Manu, who whole-heartedly supported this important initiative. It was an honor and a privilege for me to serve as President of AASIO from January 2003 to December 2006.

Manjit S. Kang

Sustainability of Agriculture: Issues, Observations and Outlook

C. Jerry Nelson

SUMMARY. Sustainability of agriculture is based on economic returns, environmental preservation, and sociological factors associated with quality of life. Yet, its components, except for economic return, are difficult to quantify, and their relative importance varies among economic and geographic regions. Environment-friendly agriculture adds emphasis to environmental factors and eco-friendly agriculture adds biodiversity to the environmental component. But sociological factors are very site-specific and hard to measure and value. Internationalization of agriculture puts pressure on sustainability because world trade is based on economics, and the role of private industry and the utility of most agricultural research are perceived to favor the economic component. Local people and national governments add emphasis to the environment and some to the social factors. As national economies improve, however, more emphasis by the public and national governments is placed on the environment and grows to include biodiversity. But actual biodiversity tends to decrease as incomes grow, due partly to larger fields and more mechanization. The combined effects of global climate change, WTO-GATT negotiations, and availability of energy and water will strongly influence crops, cropping systems, world agriculture and sustainability. Diets are changing, especially in Asia with reduced dependence on rice. And, expectations by the global public regarding the

C. Jerry Nelson is affiliated with the Agronomy Department, 108 Waters Hall, University of Missouri, Columbia, MO 65211 USA (E-mail: nelsoncj@missouri.edu).

[Haworth co-indexing entry note]: "Sustainability of Agriculture: Issues, Observations and Outlook." Nelson, C. Jerry. Co-published simultaneously in *Journal of Crop Improvement* (Haworth Food & Agricultural Products Press, an imprint of The Haworth Press, Inc.) Vol. 19, No. 1/2 (#37/38), 2007, pp. 1-24; and: *Agricultural and Environmental Sustainability: Considerations for the Future* (ed: Manjit S. Kang) Haworth Food & Agricultural Products Press, an imprint of The Haworth Press, Inc., 2007, pp. 1-24. Single or multiple copies of this article are available for a fee from The Haworth Document Delivery Service [1-800-HAWORTH, 9:00 a.m. - 5:00 p.m. (EST). E-mail address: docdelivery@haworthpress.com].

environment, biodiversity, food quality, food safety, and other multi-functional outputs that enhance quality of life are expanding the meaning of sustainability and presenting a new paradigm for agriculture. Biotechnology and precision-agriculture offer new potentials, mainly in countries that honor and protect patents. New public policies will emerge to regulate development and use of technologies to insure they are consistent with the major role of agriculturalists, to produce food in ways that assure sustainability with defined emphasis on each of its components. doi:10.1300/J411v19n01_01 *[Article copies available for a fee from The Haworth Document Delivery Service: 1-800-HAWORTH. E-mail address: <docdelivery@haworthpress.com> Website: <http://www.HaworthPress.com> © 2007 by The Haworth Press, Inc. All rights reserved.]*

KEYWORDS. Sustainability, environment-friendly, eco-friendly, biodiversity, global climate change, biotechnology, precision-agriculture, WTO-GATT, water, energy

It is appropriate that sustainable agriculture be the focus of a symposium on technologies involving agricultural crop production, the environment and food security. While agriculture is an old activity, there have been rapid changes recently in development of new technologies and the manner in which crops are produced. Economic and social factors, such as world markets, the expanded roles of private industry, strong public emphasis on food quality and safety, and much greater environmental and societal expectations, are involved. Agriculturalists and producers are recognized worldwide as the primary managers and caretakers of the landscape. In the immediate future, linkages between sustainable agriculture and new technologies in developing and, especially, in developed countries will be necessary as emphasis shifts from increasing production to sustain life to distinguishing value-added traits for quality of life. This will include preservation or enhancement of the environment and biodiversity.

My goal is to address some key issues regarding sustainability at the international, national, and local levels. The perspective has been gained through extensive travels to Asia, Africa, and Europe and working with several international colleagues in the USA and in their homelands. While it is largely an overview, the purpose is to form a framework or reference point for further understanding and decision-making regarding the management of land resources in a way that maintains or increases crop production, respects local cultural and social values, and maintains or improves the environment and biodiversity. But the rules have also changed due to world trade. Today, local goals for agriculture must be considered in an international context that increases the complexity and adds both regulations and safety nets.

THE MEANING OF SUSTAINABLE AGRICULTURE

Terminology describing a multi-purpose agriculture can be both polarizing and compromising. The term *sustainable agriculture* has been used to describe systems and cropping strategies that are economically viable, environmentally sound and socially acceptable (Figure 1), with a goal of intergenerational equality by meeting the needs of the present while leaving equal or better opportunities for the future. Nature and society are integral parts of the system. This is a relatively broad definition that implies holistic management systems to optimize a balance of outputs among economic returns, environmental preservation and social benefits. Yet, both inputs and outputs of social and environmental importance are hard to quantify or assign values leaving the impression that the primary interest of today's agriculture is economic return.

To date, there has been no universal base or common denominator, such as currency, to allow quantification of the environmental or social values that permit direct comparison with economic return. Despite this dilemma, the land manager and the policy maker are confronted by public demands to consider more than the production (economic) component, predicated largely by the inherent responsibility farmers and ranchers have for maintenance of the rural landscape and its environment. Rural-urban conflict can arise when public concern, be it perception or reality, about environmental and social aspects is ignored or under-valued relative to economic aspects (CAST, 2002).

Much of the public assumes organic agriculture is sustainable, but that is not always the case and cannot be assured. *Organic agriculture* considers mainly the source or type of inputs and their judicious use, does not allow chemical fertilizers or pesticides, and precludes use of foreign materials, such as DNA via biotechnology (AMS@USDA, 2005). The strict requirements for crops to be certified as "organic" places great pressures on allowable inputs and nearly always leads to lower productivity that needs to be offset by market price. Organic agriculture has implied concern about the environment and biodiversity. But, aside from pesticide testing of the product, there is no guarantee that chemical fertilizers were not used or that the food is safer. It is well

FIGURE 1. Left. Conceptual diagram showing the three components of sustainable agriculture with equal emphasis and some external factors that affect each component. Right. The way the distribution of outputs is perceived depends on the individual and his understanding, but consistently suggests that the economic benefit is both the primary driver and benefactor of agriculture. The ideal distribution is very site and evaluator specific and, due to environmental fragility and cultural perspectives, would need to be quantified for local conditions.

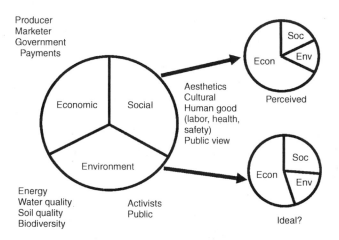

known the nutrients from crop residues and manures are not distinguishable from chemical fertilizers in the product. Yet, organic food production is the fastest growing agriculture industry in much of the developed world, mainly due to concerns with pesticides and interests in the environment and food safety. Largely households with high incomes drive this market.

In contrast with organic agriculture, *environment-friendly agriculture* allows chemical and biotechnology inputs that are either utilized or stored in the cropping system in such a way that food safety is not compromised and the environment is not altered negatively or, usually, is improved. The environment focuses mainly on quality of soil, air and water, but in some cases includes odors or aesthetic values that are not easy to define or quantify. Inputs can be synergistic; for example, fertilized plants that are vigorous use nutrients more efficiently and need fewer pesticides. Adding biotechnology can further reduce use of pesticides or allow use of pesticides that are safer for the public and the environment (CAST, 2005).

Eco-friendly agriculture is more inclusive and considers the environment in concert with the total ecosystem, including its biodiversity and function (Lemaire et al., 2005). Indirectly, this goal adds emphasis to the environmental component and, especially, to the social dimension of sustainability. But, inclusion of biodiversity in the output evaluation adds considerable complexity to the management system and requires measurements of factors that constitute quality of life. To-date most of the research on eco-friendly systems has used large scale modeling of a region or watershed, making it difficult to scale back to give priorities or guidelines to individual producers or landowners. Adoption of this approach to crop management on a large area, perhaps even a watershed, will require mutual agreement among several adjacent landowners or managers. It will also involve the public to a great extent and desired outputs will likely differ from area to area.

In many ways, environment-friendly agriculture embraces the principles of sustainable agriculture with added emphasis on the environmental component. This approach can utilize more modern technologies to maintain productivity and can enhance both the economic and environmental aspects (Cassman et al., 2003). Most conclude that chemical fertilizers need to be used to support productivity and that biotechnology offers great promise for crop improvement. However, in many ways environment-friendly agriculture will need governmental intervention and payments to farmers. Increases in production alone, especially with high land costs, will not provide the income needed for the farmer or landowner to cover the costs to maintain the environmental and sociological aspects of sustainability. This dichotomy must be addressed to accomplish the entire package, in that farmer/producer decision-making is primarily driven by economics whereas public decision-making is driven primarily by expectations concerning environmental and social issues. The public will need to help fund these broader aspects.

While the concept of environment-friendly agriculture is highly desirable, it is not without its challenges from the organic-minded populace, the need for development and use of modern technologies, and the concerns of those who feel biodiversity is slighted. Scores of research reports have clearly indicated that the world today cannot be fed totally with organic production, and most conclude that fertilizer use is essential, especially nitrogen for cereals (Cassman et al., 2003). In addition, while using more inputs and technologies than organic agriculture, the production and marketing systems for environment-friendly agriculture continue to recognize the public demand for an adequate and safe food supply. This seems to be the logical goal while honoring and embracing cultural aspects and perspectives on biodiversity.

ECONOMIC DEVELOPMENT
AND ENVIRONMENTAL PRESERVATION

Often it is assumed that strong economic growth will be at the expense of the environment, but that is usually not the case (Ekins, 1999). Recently that concept has been evaluated using an Environmental Kuznets Curve (Harbaugh et al., 2002; Nahman and Antrobus, 2005) with good sources of data that show that the environment generally improves with economic growth (Figure 2). In this data set, the Environmental Stability Index is a composite of 13 environmental quality indicators that include air quality, water quality, and environmental health, each given equal weight. The data presented are largely for developing countries and a few are labeled. The world average is about 50 with Finland, Norway, Uruguay, Sweden and Iceland having the highest scores and North Korea, Iraq, Taiwan, Turkmenistan, and Uzbekistan having the lowest scores. Yet less than 25% of the total variance is associated with economic level, suggesting other factors, e.g., education levels or government priorities of high-income countries, also have an influence.

It is generally concluded some environmental degradation will occur during early stages of economic development even though there is a small rise of per capita income. But as incomes continue to rise, there is an increased demand for improvements in the environment and other non-economic factors affecting quality of life. In some cases, a developing country may be below expectation in environmental improvement due to rapid economic growth that precedes the implementation of governmental policies and the emergence of strong public concern.

When biodiversity issues are also related to economic status of the country (Figure 3), the relationship becomes negative and inverse from that considering only the environment (Figure 2). The Biodiversity Index includes several measures, including threats to endangered species and changes in land use. As mentioned above, although somewhat linked, maintenance of biodiversity is different from maintenance of the environment, and use of an agricultural technology may not be beneficial to both. Secondly, the importance of biodiversity to the public is more futuristic than is the importance of the environment, and is emphasized mainly in the most developed countries. Most developed countries

FIGURE 2. Effect of gross domestic product (GDP) on the index of environmental sustainability. Data points are labeled for some representative countries. The index includes several measures of soil, air and water quality. Adapted from Lee and Chung (2005).

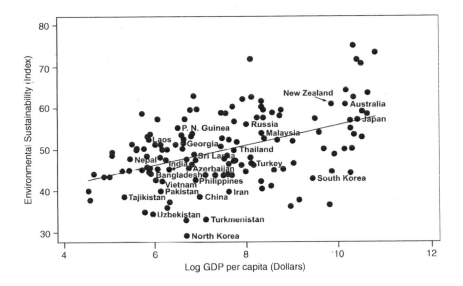

have had major reductions in their rural populace such that fewer are involved directly in commercial agriculture, leading to larger field sizes, proportionate increases in monocultures, and more dependence on mechanization. All are known to have negative effects on plant, animal and microbe biodiversity.

In developing countries, which have very low gross domestic products (GDPs), the general public will usually tolerate a short-term decrease in environmental quality for improved production, but once the rudimentary food needs are met and GDP grows, the emphasis quickly changes to an environmental priority as well as maintaining production. But in the long term, as the economy continues to improve, there is a shift in public concern from that for the environment to that of the entire ecosystem. The overall result is that, when the two relationships (Figures 2 and 3) are combined across a broad range of GDPs, the relationship with environmental improvement and change in biodiversity forms a characteristic inverse-U shape (the Kuznets curve)–first defined in 1991 by Grossman and Krueger (1995). This gives an apparent "optimization" at a middle range of GDPs.

FIGURE 3. Effect of gross domestic product (GDP) on the index of ecosystem biodiversity. Data are labeled for some representative countries. The index presented includes the National Biodiversity Index for each country and the proportion of bird, mammal, and amphibian species that are considered threatened. Adapted from Lee and Chung (2005).

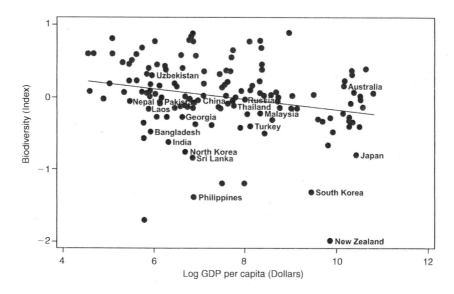

Thus, as economic growth occurs in a developing country, public expectation shifts from adoption of technology focused mainly on production to that which also improves the environment, and eventually to that which also maintains or enhances biodiversity. This is already evident in Europe where public interest and agricultural research at the ecosystem level are gaining strong momentum with minimal interest in increasing production. The smaller farm size in Europe compared with other developed countries with high per capita income probably contributes to maintaining biodiversity.

DEALING WITH GLOBAL CLIMATE CHANGE

Since the time of the industrial revolution, the CO_2 concentration of the air has increased from about 280 ppm in 1890 to about 370 ppm today. Data suggest that about 2/3 of the increase is due to the burning of fossil fuels and about 1/3 is due to changes in land use and cultivation. The latter was associated with shifts from natural ecosystems to agricultural ecosystems that were subject to increased soil erosion and rapid decrease in soil organic matter (Lal et al., 2005).

If air temperatures would remain similar, most C_3 crops would have reduced photorespiration at the higher CO_2 concentration, about 30% higher productivity and increased water-use efficiency (Kimball et al., 2002). Temperatures during the past 100 years are the warmest on record, however, and the rate of change during the past 50 years has accelerated, giving credence to the direct effects of man's intervention (Rosenzweig and Hillel, 2005). The global models now suggest that by the year 2100, global temperatures will increase by a small amount near the equator to as much as $6°$ C at the poles. The accompanying temperature increase will largely offset the positive effects of CO_2 by increased respiration and photorespiration of C_3 plants. The melting of the polar ice with concomitant increase in the sea level will lead to more floods and even inundation of some land currently used in agriculture.

In addition to higher CO_2 and other greenhouse gasses, there will be more heat stress and transpiration (Figure 4). C_4 crops use water more efficiently and will be less affected than C_3 crops grown in drought–prone environments, yet heat stress will further exacerbate the drought problems for both. Disease organisms and insect pests will have more rapid life cycles so there will be greater needs for pesticides or for better genetic resistance through conventional breeding or biotechnology. Annual rainfall is expected to increase slightly, but there will likely be more severe storms and increased soil erosion (Figure 4). The growing season for crops will be longer, which may allow more areas to have double or relay cropping systems. There could be increased need for winter cover crops to protect the soil, but more soil compaction will occur due to increased animal and/or wheel traffic in the fields.

Technologies to offset global change will be focused on favorable crop sequencing, water conservation, soil conservation, soil temperature control, and water-use efficiency. Short-term emphasis is likely going to be on environmental conservation, especially soil erosion and capacity for carbon sequestration in the soil, but ultimately will need to include soil compaction and biodiversity. Use of technologies like precision agriculture and biotechnology will offer great potential for improving crop management and the environment by increasing tolerance to biotic and abiotic stresses. In the short term, more CO_2 will need to be captured in the soil organic matter, which will depend to a great degree

FIGURE 4. Possible benefits and drawbacks from climate change mediated by an increase in CO_2 and other greenhouse gases. The longer growing season and faster rate of development will allow more than one crop to be grown annually, but likely with more pest and disease problems. The increased precipitation and more severe storms may alter crop production, especially in monsoon climates and areas subject to soil erosion. Adapted from Rosenzweig and Hillel (2005).

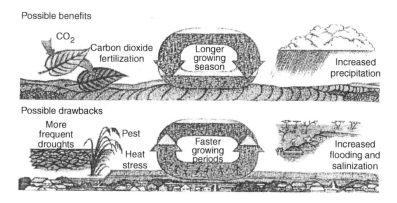

on technologies to allow reduced tillage and other means of soil management (Figure 5).

Unquestionably, global change will cause compromises for crops and cropping systems, including continuing to increase production on the best producing and least erosive soils. If policies and economics at the international and local levels are not conducive, there will be more pressures to use lower quality land sites for production, which will cause more social and environmental concerns. Agriculturalists around the world will need to use all the technologies available to protect the best soils to provide the needed food and other components. This will also be a great concern in developing countries as world markets develop as they will need support from developing countries to help offset the disadvantages they will have in world trade.

THE BIOTECHNOLOGY MOVEMENT

Researchers and farmers have worked together for several decades to develop, learn and adopt new technology (Figure 6). The progress through this synergism grew stepwise at a gradual rate. As plant breed-

FIGURE 5. Effects of soil texture (actually percent clay in each class) and aggregation potential on carbon sequestration in soils. More organic matter can be accumulated in heavier soils to support aggregation. The rate and extent of carbon accumulation will be strongly affected by aggregation level and tillage practices. Inset shows the effect of tillage practices on sequestering organic carbon as soil organic matter (SOM). More intensive tillage systems could decrease the organic carbon concentration. Adapted from Duxbury (2005).

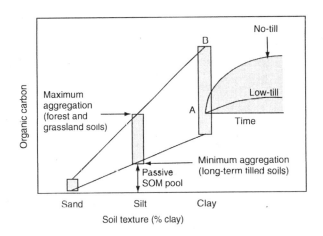

ers developed new cultivars, they could be managed more effectively, thereby allowing the next step in breeding to be made and new management practices developed (Evans, 1993). The research information was credible and reliable, and through education the farmers adopted the technology. During the last half century, however, there has been a growing role for private industry, a new partner in the effort, especially in developed countries. This partnership developed rapidly during the last few years and markedly accelerated the rate of technology development, such that most of the products or technologies in developed countries have been the result of private industry.

The relationships have also changed. First, public concerns about credibility and motives arose as the research and development of agricultural products was gradually assumed by private industry that lacked a long cultural history with agriculture. Public trust became an issue as use of chemical inputs increased and concerns were that food safety and quality might be disregarded due to profit motives. Secondly, would the sincere environmental commitment of the farmer be compromised or diluted by economic pressures of the new "chemical-based" agricul-

FIGURE 6. Crop improvement traditionally involved close working relationships between public plant breeders and practical agronomists to develop a product and its application. Now, in developed countries private industry plays an ever-increasing role with both the public plant geneticist and the agronomist to develop the cultivars and management strategies for the product.

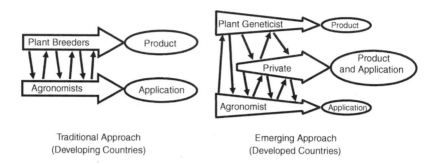

Traditional Approach
(Developing Countries)

Emerging Approach
(Developed Countries)

ture? The social and ethical issues of sustainability and food safety needed to be addressed. Unfortunately, the knowledge gained by social science research could not keep up with the technology revolution, mistakes were made in public relations and trust waned further.

Governments became more involved and stronger regulations were put in place to demand clear scientific evidence and assurance to the public regarding food safety and environmental aspects before pesticides and fertilizers could be used. Industry adjusted, and in developed countries private industry played a stronger and stronger role in agricultural development while publicly funded research on many commodities subsided. When a profit could be made from sale of a product, such as seed, pesticides or machinery, the role of industry usually took the lead in product discovery, development, and even education (Figure 6).

This shift has forced public-funded research to coincide with potentials for private industry, especially in the discovery phase and then later in the adoption and education phases (Figure 7). But industry uses the discovery to develop a product in-between and eventually controls the majority of patents or intellectual property rights on the technology. The focus of the private sector is on major markets and technologies that can be protected and are profitable. Public sector funding is used for more minor crops where markets are smaller, for addressing environmental problems, and protecting and managing natural resources, such as air, water, soil, and germplasm.

FIGURE 7. Private industry has emerged as the major contributor for develop-ing new pesticides or crop varieties of major crops in developing countries. Even some of the basic research by publicly supported scientists is patented before selling the rights to private industry. Private research then develops the product and the educational program to market the product. Applied research by publicly funded scientists may help test the product or practice in a specific region or environment and provide an independent assessment.

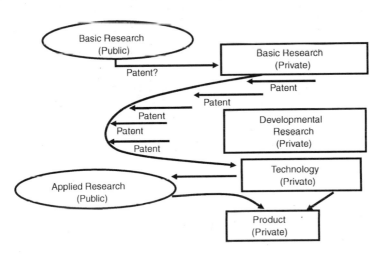

Whether technology is developed and made available through public agencies or through private industry often depends on the potential for patenting and protection of intellectual property. Once a patent is granted, a private company can determine the financial risk and bene-fit of developing the technology to produce and market the product (Figure 7). This is relatively straightforward for technologies such as a new pesticide or a piece of farm machinery that can be mass-produced to capitalize on economy of scale and sold through a defined market. Farmers cannot afford to manufacture their own pesticide or build their own piece of equipment. It is cost effective to use the product so they purchase it, which gives both the manufacturer and user a direct benefit. This seemingly win-win situation is similar for improved crop cultivars.

Biotechnology in cropping systems came of age in the late 1990s and early 2000s, mainly through the use of Bt genes in several crops to give excellent resistance to some species of insects and the incorporation of glyphosate (Roundup) resistance into maize and soybean (Fernandez-Cornejo and Caswell, 2006). Farmers in some countries quickly adopted both, while governments in other developed and developing countries prohibited production or import of any genetically engineered crop (Hoban, 2004). Concerns were generally raised about food safety, the environment, and threats to biodiversity. These perspectives likely were tempered by the role and distrust of industry.

More farmers are choosing to plant more genetically modified (GM) crops than ever before because of the real advantages and benefits. Fortunately, a growing number of countries allowed use of biotech or GM crops in 2005, including five European Union countries. Globally, the trend is sharply upward with 67, 81, and 90 million hectares of transgenic crops being produced, respectively, in 2003, 2004 and 2005 (James, 2005). In order, most was produced in the United States, Argentina, Brazil, Canada, and China, while India had the highest rate of adoption. Biotech traits were still mainly insect resistance and herbicide tolerance in soybean, cotton and maize. Farmers using GM crops increased their income by US$27 billion during the period 1996 to 2004, along with significant environmental benefits. Of great importance is that developing countries benefited more ($15 billion) than developed countries ($12 billion).

The general public has less confidence in the private sector regarding biotechnology, especially in Europe and parts of Asia (Hoban, 2004). Yet studies show that economic returns from Bt cotton in the USA were 20-45% to farmers, 36-38% to the industry, and 19-27% to consumers (Traxler, 2004). With herbicide-tolerant (glyphosate) soybean, the returns were 13% for farmers, 34% for industry, and 53% for consumers, due largely to greater area of production and lower prices for the products. It is evident that industry is not the only benefactor of biotechnology, and the overall result is actually a "win-win-win" situation (Fernandez-Cornejo and Caswell, 2006).

Biotechnology-derived crops in adapted areas have provided large economic and environmental benefits, partly due to increased yield, but mainly by reduced use of pesticides and improved efficacy of those that are used. These technologies also helped reduce needs for tillage and added flexibility for timing of needed pesticide applications. And the environmental benefits are enormous due to the marked reduction of insecticide use in cotton and the use of glyphosate herbicide on soybean and other crops instead of more toxic and persistent herbicides. In addition, the reduced use of tillage for soybean and cotton crops has reduced energy consumption, left water-saving residue on the soil surface, and has led to reduced soil erosion and the associated benefits for water quality. Traxler (2004) concluded that use of Bt has likely been the greatest economic technology ever developed for cotton.

According to the above, most GM crops are environment-friendly and have improved economic conditions for producers and for private industry, the other major economic benefactor. Unfortunately, due to social concerns, patents and higher seed costs, GM crops have not been available to farmers in many developing countries and in some of the most developed countries, which is an interesting dichotomy. Use of biotechnology will be slow in developing countries unless necessary patent and intellectual property protection is in place to protect private investment. This is brought out in the Roundup™-ready soybean situation in Argentina that allowed biotechnology crops, whereas adjacent Brazil had policies to prohibit the same crops. Soon it was discovered that much of the Brazilian soybean crop included the Roundup-ready gene that had been smuggled in as seed from Argentina. The technology could easily be utilized in the self-pollinated crop, although it was illegal according to both Brazilian law and patent rights of Monsanto Corporation. Fortunately, most developing countries are addressing the patent issues.

Issues regarding food safety and potential effects on biodiversity also reduce use of biotechnology, especially as per capita income rises. A lack of education of the public in some developing countries will further delay use of biotechnology as the populace has difficulty evaluating the science associated with food safety and biodiversity implications. Having conflicting policies among the developed nations does not improve the situation and, in countries with low education levels, action groups can have a disproportional effect on public opinion and eventual government policy.

Most countries have already implemented food safety criteria and testing procedures for crops improved by non-sexual gene transfer via biotech that are much more stringent than those for crops improved via sexual crosses and conventional breeding methods. This has forced private industry to invest millions of dollars to gain approval for each product. Scientists generally agree that foods derived from biotechnology are actually much safer than those derived from conventional breeding.

Attitudes of consumers toward biotech crops differ among and within countries, and the attitudes gradually change (Hoban, 2004). Those in the USA and in Asia tend to be more favorable toward genetically engineered crops (Fernandez-Cornejo and Caswell, 2006), whereas those in Europe tend to be much less favorable. But in all countries, there are significant numbers with reservations. The exact reasons for the differences in opinion are difficult to discern, but usually involve food safety, effects on biodiversity, and general political issues mainly about having 'control of agriculture' by private industry and the concentration of those industries in a very few countries. Attitudes gradually change in both positive and negative directions, but again the reasons for the change are largely not known. For example, it was felt that the food safety issues with biotechnology crops were being resolved, the reduced pesticide use and other environmental benefits were being recognized, and lower food prices were being accomplished. But emerging interests related to growing pharmaceutical plants and biopharming raised a new higher level of concern, even in the United States, based mainly on safety factors and biodiversity issues. This public perspective is unique in that many routine pharmaceuticals are being produced using biotechnology of microorganisms.

WORLD TRADE ORGANIZATION (WTO): GENERAL AGREEMENT ON TARIFFS AND TRADE (GATT)

Rapidly approaching is full implementation of the WTO agreements. Currently efforts are being made to work out the various market classes and their criteria for each commodity and product. This will likely take time for complex products, but may develop quickly for general commodities such as maize, soybean, and established market classes of wheat. Technologies are needed for verifying quality of the commodity and each value-added factor to establish and monitor separate market channels that retain the true value of the product. Analyses using biotechnology methods can play a positive role in verification. Likely, biotechnology-derived crops will have separate channels, but details need to be worked out.

Value-added traits, such as those in golden rice, may command a different price on the global market than conventional food-grade rice, and rice of a certain starch composition may have enough difference in quality to demand a higher global market price. Post-harvest technologies associated with preserving the value-added trait during processing, transporting, and storage will be important to the global consumer. These technologies will likely be industry-driven and involve food scientists and marketing specialists suggesting an advantage for developed countries to capture these markets.

In addition, there will be marketing opportunities for other value-added designations, such as 'organic,' 'sustainable,' 'environment-

friendly,' or 'eco-friendly' although they may not have the same importance or perceived value in an international marketing system as in local
systems. Definitions and guidelines exist for international marketing of
organic products (NAL-USDA, 2005), but will need to be established
for other designations. Practices that benefit the environment or biodiversity at a given location will be of great value to the local and perhaps
regional populace, but are less likely to be perceived as a value-added
trait further away and especially in international markets. Conversely,
the perceived health or safety benefits of purchasing food crops that
are produced organically will likely translate to those same diverse
consumers. Many developing countries could benefit from marketing
value-added crops if the marketing structure and regulations are in
place or can be developed.

THE CHANGING DIET AS COUNTRIES DEVELOP

A major consideration for sustainability is the relationship among
global climate change, world trade, biotechnology and altered diets of
countries as they become more affluent. Already, rice consumption in
Japan, Taiwan and South Korea is decreasing as wheat, potato, and
other caloric sources become preferred (Figure 8). China's per capita
consumption of rice is at a plateau and will likely decrease as incomes
rise. India also appears to show a reduction in consumption. Since the
population is still increasing in China, there will be a need for more production, but when per capita consumption decreases faster than the population increases, total need for rice is expected to decrease beginning in
about 2030. Of parallel interest is the increased demand for meat and
meat products as economies improve, a characteristic of many Asian
countries, including China. This adds pressures on the land resource to
produce feed grains as commodities compared with rice for food. Due
to cultural and religious reasons, expectations for shifting to meat consumption in India are somewhat different from China because the diet is
predominately vegetarian.

How global climate change will alter the monsoon climate is of great
concern. Rice is one of the few crops that can be produced in these areas
of high seasonal rainfall, which raises questions about suitable alternatives for the traditional rice paddies, both culturally for the public and
economically for the rice farmers. Already, as much as 1/3 of the
long-term rice paddies in Japan, Korea, and Taiwan are currently used

FIGURE 8. Per capita consumption of rice in several Asian countries (US Department of Agriculture, 2005). As economies improved the per capita consumption of rice decreased, first in Japan in the mid-1960s (−1.8%), followed by Taiwan about 1970 (−3.4%), South Korea in the late 1970s and China in 1991. India also appears to be decreasing. Total need for rice in each country will decrease when the increase in population no longer exceeds the decrease in per capita consumption. But the decrease-rate for China is projected to be only −0.33% so production will be maximum about 2020. Most Asian societies are expected to follow the trend to more consumption of wheat and meat that depends on feed grains such as maize and sorghum.

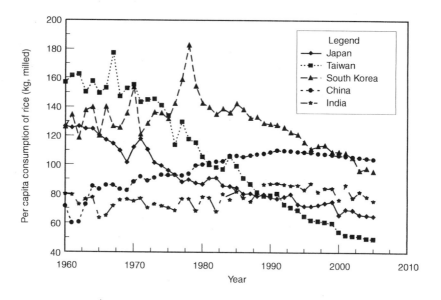

to produce other crops, but this requires educational programs, markets and new production practices for farmers. At the same time, this is a great opportunity to shift to high-value crops with a market, perhaps even international.

Unfortunately, few crops can tolerate the monsoons like rice. How will this fit together into the scheme of local and global markets? If it remains economical to produce rice, the amount not needed as a reserve for food security could be used for animal feed or for feedstocks to produce fuel or other products. Several developing countries in Asia will continue to be heavily dependent on rice as a food crop for the next few decades, but will likely gradually follow the trend.

WATER, ENERGY, AND BIODIVERSITY

Emerging rapidly are conflicts and potential solutions on the usages of water and energy resources for agriculture and other public uses. This is also linked to the desire for biodiversity, which becomes an important component of managed and natural ecosystems as income levels increase and environmental problems are seriously addressed. Water is a universal issue as agriculture is a major user; from 70-80% of the current water usage is in agriculture, largely for irrigation. Energy is already an international problem, and although new technologies are emerging to supplement fossil fuels, they are not expected to be major sources in the near future. Emphasis on biodiversity is clearly evident in Europe, with the growing recognition and priority in other countries that this issue needs to be addressed as a component of agriculture.

Biofuel resources can contribute extensively to help with the energy shortage, especially in developed countries that have excess capacity for production of crops with fuel potential, such as maize or soybean. Consistent with concerns about what crops to grow in the monsoon areas may encourage even rice to be used as a fuel substrate in Asian countries. Here, biotechnology may be a major contributor as crops can be improved to use less water, be flooding tolerant or be more efficient in biomass conversion to fuel sources.

New conservation methodologies, such as reduced tillage, or developing early maturing crops that can be field dried with low losses due to weather or pests, will help reduce inputs of fossil fuels. Unfortunately, energy sources have increased markedly in price, which will delay use of agricultural technologies for energy use in many lower-income countries, further placing them at a disadvantage. Alternative energy sources from biomass that are operated on a small scale may be of benefit to the small, subsistence farmer. Of concern, however, is the potential for redirecting manures and plant biomass to energy at the expense of their use for good soil management to provide nutrients and organic matter for sustainability.

Technological advances in agriculture are often perceived to be detrimental to biodiversity (Figure 3) and energy-use efficiency because larger fields of monocultures disturb diversity of the natural environment and promote more mechanization. Other advances in pesticides or fertilizers also require energy for their manufacture and application, and tend to cause shifts toward more intensive land use. With energy prices increasing, the energy budgets to produce a crop relative to the energy captured in the crop product are being questioned. Energy use as a measure of sustainability of agriculture will be a much higher priority in the future.

As income levels rise, there is more competition for use of the limited water resources, and agriculture is generally considered to reduce water quality rather than improve it. Conflicts on water use have occurred for a long time and there are international agreements recommending integrating the water needs for agriculture and the natural ecosystems. Improvements in cultural practices, such as tillage, residue management, weed control and more strategic times of irrigation, have increased water-use efficiency of crop production. There will be continued efforts by plant breeders and cropping systems specialists to improve the efficiency, but major changes will likely require biotechnology or a shift to greater use of drought-resistant crops.

It is already evident that agriculture will be reduced in priority as water needs for urban areas increase. Desalinization of sea-water may be a technology to provide water for public uses as most large cities and population centers are located within 75 km of the sea. There is a great interest in use of agriculture to produce biofuels and industrial feedstock that would reduce dependence on petroleum sources. In addition, there will be a need for continued efforts to decrease energy and water use on a global and local basis. At the same time, there will be needs to increase biodiversity, mainly in developed countries, but it is gradually becoming a higher priority issue in developing countries. This is seemingly a great challenge as the methodologies for measuring and understanding the intrinsic value of biodiversity are not yet clear. New technologies, likely involving social scientists and ecologists, will be necessary.

PRECISION OR SITE-SPECIFIC AGRICULTURE

The rapid development of precision agriculture will be a good foundation for dealing with issues of sustainability. To-date, precision agriculture has been studied and applied to deal with in-field variation in terms of crop management based on soil properties and treating localized areas affected by biotic stresses from weeds, insects and diseases. In the USA, the early adopters used yield monitors to evaluate potentials for dealing with the inherent spatial variability of the soil, and have learned that soil hydrology is highly variable and is likely the major factor regulating crop growth (Hamza and Anderson, 2005).

Adjusting soil fertility applications based on soil tests and targeting localized areas for pesticides have decreased the potentials for pollutants entering the environment, but may not be economic. In general, conventional soil sampling and testing are often not precise enough to justify application of specific rates of fertilizers (Ruffo et al., 2005). This has increased the interest in tissue analysis and remote systems of tissue testing, such as chlorophyll meters and other non-destructive instruments at specific growth stages, especially for nitrogen (Fox et al., 1994; Zhao et al., 2003; Belanger et al., 2005).

An expanded approach to precision agriculture may be useful for understanding the relationships of soil properties, water use, and energy needs for integrating crop production, environmental conservation, and biodiversity across a watershed or landscape. Use of appropriate models and good data on soils and cropping systems should predict a diverse landscape of crops, trees, grasslands, and livestock in a spatial pattern that optimizes land use to improve the environment and biodiversity while maintaining high crop productivity. New technologies in spatial statistics should assist in determining where each component in the overall system should be located in the landscape. For example, in a Missouri watershed, switching crops from fields that are located near to areas more distant from a stream had a major effect on the water quality. But cropping arrangements such as these will require cooperation of the several farmers or land owners that occupy the total area.

Except in socialist countries, farmers tend to independently make assessments and set goals, whereas managing a watershed for diversity and environmental aspects of sustainability requires collective efforts. Accomplishing such a goal will require government encouragement and perhaps intervention, which will likely be decided by public policy and only partially by the farmers or landowners. For a long time, the farmer or landowner could decide how to use the land, but today the social environment has changed markedly. Land is now considered to be a public resource and the citizenry has assumed a major role in determining how it is used and managed. Food security and a good environment are universal priorities. As agriculturalists, we can learn a lot from the automobile industry as it has been forced by the public and government intervention to focus on safety issues and a healthy environment.

Many countries are investing heavily in use of crops or animal wastes as biofuels and the use of wind energy sources to generate electricity. Clearly, there are advantages and disadvantages to uses of these renewable or sustainable sources of energy. And soil and crop management strategies will need to help sequester CO_2 that continues to be produced by burning fossil fuels. But again, the best soils for CO_2 storage may be site specific (Figure 5), and crop species and management strategies will need to be designed. New technologies will also be needed in tillage systems, irrigation, fertilizer production, and pesticide production to reduce the energy needs of agriculture. Similarly, new technologies will be needed to reduce energy costs for mechanical seeding, application of agrichemicals, and harvest, especially those that also reduce soil compaction from wheel or foot traffic (Hanza and Anderson, 2005).

CONCLUSION

New technologies are being applied to address sustainability and to improve environment-friendly agriculture even though the problems are more complex to address and more difficult to solve than for production alone. Nevertheless, agriculture is positioned to continue to provide leadership for the effort, even including the emerging issues of biodiversity. Agriculture within an ecosystem is linked in resource use, biological processes, and research methodology. So it is natural that increasing food production can be achieved while improving the environment and ecosystem. Even in developing countries, there can be win-win situations for food production and environmental services, especially if private industry can assist and market sources are available for economic benefit to the land managers. How that economic benefit is generated and who are the benefactors will be topics of international discussion to determine if and how government subsidies are used to give a fair world marketplace and sustainability to the overall system.

Three basic issues, namely *knowledge, values,* and *institutions,* are important and integral to decisions to harmonize crop production and the ecosystem. Broader and more specific knowledge about systems is needed along with the appropriate technologies to change or improve the conditions. Values associated with the environment, social well-being and quality of life are harder to define, and will require assistance from social scientists to quantify and interpret. Yet this component of sustainability is absolutely critical as societal values and perceptions will determine which technologies the public will allow agriculturalists to use. And institutions will have a major role, especially in terms of integrating public and private research. Institutions will also be responsible for educational programs for the producer to aid in adoption of new technologies, and to the public about the positive attributes of agriculture and crop production that are consistent with improvement of the environment, ecosystem and quality of life. Fortunately, the agricultural community is recognizing its role and is preparing to address the problems.

REFERENCES

AMS@USDA. (http://www.ams.usda.gov/nop/NOP/standards.html) Verified 10 Dec. 2005.

Belanger, M-C., A.A. Viau, G. Samson, and M. Chamberland. 2005. Determination of a multivariate indicator of nitrogen imbalance (MINI) in potato using reflectance and fluorescence spectroscopy. Agron. J. 97:1515-1523.

Cassman, K.G., A. Dobermann, D.T. Walters, and H. Yang. 2003. Meeting cereal demand while protecting natural resources and improving environmental quality. Annu. Rev. Environ. Resour. 28:315-358.

CAST (Council for Agricultural Science and Technology). 2002. Urban and agricultural communities: Opportunities for common ground (R138). CAST. Ames, IA. 124 p.

CAST (Council for Agricultural Science and Technology). 2005. Crop biotechnology and the future of food: A scientific assessment. CAST Commentary QTA 2005-2. CAST. Ames, IA. 6 p.

Duxbury, J.M. 2005. Reducing greenhouse warming potential by carbon sequestration in soils: Opportunities, limits, and tradeoffs. Pp. 435-450. *In* R. Lal et al. (ed.) Climate Change and Global Food Security. CRC Press, Taylor and Francis Group, Boca Raton, FL.

Ekins, P. 1999. Economic Growth and Environmental Stewardship: The Prospects for Green Growth. Routledge Press, London.

Esty, D.C., M.A. Levy, T. Srebotnjak, and A. de Sherbinin. 2005. Environmental Sustainability Index: Benchmarking National Environmental Stewardship. Yale Center for Environmental Law and Policy. New Haven, CT, USA.

Fernandez-Cornejo, J., and M. Caswell. 2006. The first decade of genetically engineered crops in the United States. USDA-ERS Economic Information Bulletin 11. http://www.ers.usda.gov/publications/E1B11/ (accessed on June 15, 2006).

Fox, R.H., W.P. Piekielek, and K.M. Macneal. 1994. Using a chlorophyll meter to predict nitrogen-fertilizer needs of winter-wheat. Comm. Soil Sci. Plant Anal. 25:171-181.

Grossman, G.M., and A.B. Krueger. 1995. Economic growth and the environment. Quart. J. Econ. 110:353-377.

Hamza, M.A., and W.K. Anderson. 2005. Soil compaction in cropping systems: A review of the nature, causes and possible solutions. Soil & Tillage Res. 82:121-145.

Harbaugh, W., A. Levinson, and D.M. Wilson. 2002. Reexamining the empirical evidence for an environmental kuznets curve. Rev. Econ. Stat. 84: 541-551.

Hoban, T.J. 2004. Public attitudes towards agricultural biotechnology. ESA Working Paper 04-09 (www.fao.org/es/esa) (accessed on June 15, 2006).

James, C. 2005. Global status of commercialized biotech/GM crops: 2005. ISAAA Briefs 34-2005, International Service for the Acquisition of Agri-Biotech Applications (http://www.isaaa.org) (accessed on June 15, 2006).

Kimball, B., K. Kobayashi, and M. Bindi. 2002. Responses of agricultural crops to free-air CO_2 enrichment. Adv. Agron. 77:293-368.

Lal, R., N. Uphoff, B.A. Stewart, and D.O. Hansen. 2005. Climate Change and Global Food Security. CRC Press, Taylor and Francis Group, Boca Raton, FL.

Lee, H-H., and M.K. Chung. 2005. On the relationship between economic growth and environmental stability in Asia and the Pacific. Pp. 47-71. *In* Proc. 7th Truman Conference, in Memory of the 60th Anniversary of the Pacific War's End. 19 May; 2005. MU Alumni Assoc., Seoul, Korea.

Lemaire, G., R. Wilkins, and J. Hodgson. 2005. Challenges for grassland science: Managing research priorities. Agric. Ecosyst. Environ. 108:99-108.

NAL-USDA. 2005. Organic agricultural products: Marketing and trade resources. II. Regulations, laws and legislation. http://www.nal.usda.gov/afsic/AFSIC_pubs/OAP/srb0301b.htm (accessed on June 15, 2006).

Rosenzweig, C., and D. Hillel. 2005. Climate change, agriculture and stability. Pp. 243-268. *In* R. Lal et al. (ed.) Climate Change and Global Food Security. CRC Press, Taylor and Francis Group, Boca Raton, FL.

Ruffo, M.L., A. Boller, R.G. Hoeft, and D.G. Bullock. 2005. Spatial variability of the Illinois Soil Nitrogen Test: Implications for soil sampling. Agron. J. 97:1485-1492.

Traxler, G. 2004. The economic impacts of biotechnology-based technological innovations. ESA Working Paper 04-08 (www.fao.org/es/esa) (accessed on June 15, 2006).

U.S. Department of Agriculture. 2005. PSD Online, Feb. 2005. (http://www.fas.usda.gov/psd/complete_files/default.asp) (accessed on June 15, 2006).

Zhao, D., K.R. Reddy, V.G. Kakani, J.J. Read, and G.A. Carter. 2003. Corn (*Zea mays* L.) growth, leaf pigment concentration, photosynthesis and leaf hyperspectral reflectance properties as affected by nitrogen supply. Plant Soil 257:205-217.

doi:10.1300/J411v19n01_01

Agroforestry for Sustainability of Lower-Input Land-Use Systems

P. K. Ramachandran Nair

SUMMARY. Agriculture and forestry are too often treated separately, yet these two sectors are often interwoven on the landscape and share many common goals. The age-old practice of growing crops and trees together was ignored or bypassed in the single-commodity paradigm of agricultural and forestry development. Thanks to 25 years of efforts, agroforestry–the integration of multipurpose trees into farming systems–has now become a robust, science-based, integrated discipline. Agroforestry offers a unique set of opportunities for alleviating poverty and arresting land degradation, and providing ecosystem services in both low-income and industrialized nations. Improvement and exploitation of the large number of fruit trees and medicinal plants is but one of several promising opportunities for enhancing the food and nutritional security without causing the environmental hazards that are characteristic of input-intensive land-use systems. These agroforestry practices are partic-

P. K. Ramachandran Nair is affiliated with the University of Florida, Gainesville, FL 32611 USA (E-mail: pknair@ufl.edu).

The author wishes to thank S. C. Allen and D. S. Zamora for their assistance in the preparation of this manuscript.

Grant support was provided in part by USDA/IFAFS/CSREES (Initiative for Future Agricultural and Food Systems/Cooperative State Research, Education and Extension Service), through the Center for Subtropical Agroforestry, University of Florida.

[Haworth co-indexing entry note]: "Agroforestry for Sustainability of Lower-Input Land-Use Systems." Nair, P. K. Ramachandran. Co-published simultaneously in *Journal of Crop Improvement* (Haworth Food & Agricultural Products Press, an imprint of The Haworth Press, Inc.) Vol. 19, No. 1/2 (#37/38), 2007, pp. 25-47; and: *Agricultural and Environmental Sustainability: Considerations for the Future* (ed: Manjit S. Kang) Haworth Food & Agricultural Products Press, an imprint of The Haworth Press, Inc., 2007, pp. 25-47. Single or multiple copies of this article are available for a fee from The Haworth Document Delivery Service [1-800-HAWORTH, 9:00 a.m. - 5:00 p.m. (EST). E-mail address: docdelivery@haworthpress.com].

ularly suitable for resource-limited conditions and lower-input situations, which cover an estimated 1.9 billion hectares of land and 800 million people in developing countries. doi:10.1300/J411v19n01_02 *[Article copies available for a fee from The Haworth Document Delivery Service: 1-800-HAWORTH. E-mail address: <docdelivery@haworthpress.com> Website: <http://www.HaworthPress.com> © 2007 by The Haworth Press, Inc. All rights reserved.]*

KEYWORDS. Sustainability, agroforestry, developing countries, food production

INTRODUCTION

The world has witnessed dramatic increases in agricultural production during the past half century (1950 to 2000). In response to the increasing world population (that more than doubled: from 2.5 billion in 1950 to 6.1 billion in 2000) and expansion of the world economy (that increased more than seven-fold from $7 trillion in 1950–in 2001 dollars–to $46 trillion in 2000, worldwide), the world grain production tripled during this period, from 640 million tons in 1950 to 1,855 million tons in 2000. Out of this 190% increase in grain production, only 30% was the result of increases in area under cultivation, while the remaining 160% was made possible by increases in yield per unit area (world grain yield per ha increased from 1.06 tons in 1950 to 2.79 tons in 2000–and is projected to be 2.99 tons by 2010) brought about by development and adoption of modern agricultural technology. Norman Borlaug has famously articulated that this higher grain production per unit area brought about by new agricultural technologies on 660 million ha has spared 1.1 billion ha of forest land from being cleared (i.e., the area that would need to have been cleared to produce the same amount of food grains if modern technologies were not used) (http://www.usda.gov/oce/forum/speeches/borlaug.pdf).

These are impressive accomplishments, indeed. But the question that has begun to be asked in recent years is, what is the "real" cost and long-term impact of these gains? Obviously there are different scenarios and viewpoints on such a complex question. But the crucial question is: Are the modern technologies causing increasing damage to the ecological foundations of agriculture, such as land, water, forests, biodiversity, and atmosphere? In other words, in our efforts to provide for the needs of the present, are we compromising the ability of future generations to provide for themselves or are these technologies sustainable?

Lester Brown argues in his compelling book *"Plan B"* that our principal threats are now more environmental than military (Brown, 2004). He argues that our claims on earth for feeding the expanding populations have become excessive. We are asking more of the earth than it can give on an ongoing basis. Throughout history, humans have lived on the earth's sustainable yield–the interest from its natural endowment. But now we are consuming the endowment itself. Wackernagel et al. (2002) reported (in a study published by the U. S. National Academy of Sciences) that humanity's collective demands first surpassed the earth's regenerative capacity around 1980; our demands in 1999 exceeded that capacity by 20%. On the eve of the U. N. Conference on Environment and Development held at Rio de Janeiro in June 1992, the Union of Concerned Scientists published an open letter titled "World Scientists' Warning to Humanity." It stated, "Human beings and the natural world are on a collision course." The letter stated further, "If not checked, many of our current practices put at serious risk the future that we wish for human society and the plant and animal kingdoms, and may so alter the living world that it will be unable to sustain life in the manner that we know." More than 1600 scientists from leading scientific academies in 70 countries, including 104 Nobel Laureates, signed the letter.

How are ecology and environment impacted by intensive agriculture and forestry? On the agronomic front, raising land productivity has depended on using more fertilizer, expanding irrigation, and controlling pests (insects, diseases, and weeds). For example, the growth in the world fertilizer industry after WW II has been spectacular. Between 1950 and 1990, fertilizer use climbed more than ten-fold, from 14 million to 146 million tons.[1] Paralleling this was the near-tripling of irrigated area; during the early part of this period, growth in irrigation came largely from the building of dams. When the number of undeveloped dam sites diminished by the late 1960s, attention was turned to underground water, and millions of irrigation wells were drilled during the later period. About 80% of China's and 50% of India's grain harvests come from irrigated land. In the USA, irrigated land accounts for 20% of the grain harvest.[2] In many parts of the world, the need for water is simply outgrowing sustainable supply. And, vast areas of irrigated land have become nonproductive due to salinity.

Thus, we are faced with a very complex problem. We need to increase land productivity to meet the growing demands of food and fiber, for which use of nonrenewable inputs (fertilizers and pesticides) is essential–at least according to our current thinking. But we also need to reduce the use of these inputs for the sake of environmental integrity and ecological balance. It is somewhat analogous to the obesity problem that the developed nations, particularly the USA, are facing today: consuming excessive amounts of nutritiously imbalanced food leads to obesity. Just like balanced nutrition and healthy life style are the best recognized remedies for obesity, proper land care that makes the best use of nature's goods and services is essential for sustaining productivity of the land. Agroforestry is such an approach to sustainable land-use, based on the age-old practice of growing crops and trees together.

EMERGENCE OF AGROFORESTRY AS AN INTEGRATED SCIENCE AND PRACTICE

Thanks to efforts over the past 25-plus years, agroforestry has been transformed from "a practice in search of science" into a science-based practice (Nair et al., 2005). It has emerged as an integrated applied science that has demonstrated potential for addressing some of the land management and environmental problems in both developing and industrialized nations. The 1st World Congress of Agroforestry held in Orlando, Florida, June-July 2004[3] signifies this coming of age of agroforestry. A Congress Declaration affirmed by the more than 500 delegates from 82 countries states that the adoption of agroforestry over the next decade will "greatly enhance the achievement of the United Nations Millennium Development Goals" (www.un.org/millennium) by increasing household income, promoting gender equity, improving health and welfare of people, and enhancing environmental sustainability.

Simply put, agroforestry involves growing of trees with crops, and/or sometimes animals, in interacting combinations in space or time dimensions (Nair, 2001). The practice has been prevalent for many centuries in different parts of the world, especially under subsistence farming conditions (Spurgeon, 1978). It was during the late 1970s that efforts were initiated to bring these traditional practices into the realm of modern agricultural science (Bene et al., 1977; Steppler and Nair, 1987). The motivations for these initiatives included frustrations arising from the Green Revolution's failure to benefit poor farmers and those in less-productive agroecological environments, and escalating land-management problems, such as tropical deforestation, fuel-wood shortage, and soil degradation on the one hand, and increased awareness about the relevance of the age-old tree-and-crop integrated farming practices on the other hand. Although the Green Revolution model resulted in significant increases in agricultural production (Evenson and Gollin, 2003), the traditional practices of raising food crops, trees, and animals together as well as utilizing an array of products from natural woodlots, did not fit into the "single-commodity paradigm" that promoted single-species stands of crops. Serious doubts began to be expressed about the relevance of the single-commodity strategies and policies promoting them. In particular, there were concerns that the basic needs of the poorest farmers were neither being considered nor adequately addressed, especially those farming in remote rural areas or in marginal environments (Bene et al., 1977). Soon it became clear that many of the technologies that contributed to the Green Revolution were not relevant to farmers in these situations. It was recognized that a major cause of tropical deforestation was the clearing of more land to provide food and fuel-wood for the rapidly increasing populations. The search for appropriate strategies to address these problems led to serious studies of age-old practices based on combinations involving trees, crops, and livestock on the same land unit (Nair, 2001; Steppler and Nair, 1987). The inherent advantages of land-use practices involving trees, such as environmental conservation, and multiple outputs, were recognized quickly. Agroforestry thus began to be recognized, and it was institutionalized with the establishment of the International Centre for Research in Agroforestry (ICRAF)–now called the World Agroforestry Centre–in 1977, in Nairobi, Kenya. Following these developments, agroforestry was incorporated into national agricultural and forestry research agendas in many developing countries during the 1980s and 1990s (Steppler and Nair, 1987).

In the temperate region, agroforestry has had a slower evolution as a science and practice than in the tropics (Garrett et al., 2000). But faced with the environmental consequences of agricultural and forestry practices that focused on the economic bottom line, the general public in developed nations started demanding greater environmental accountability of land-use practices and the application of ecologically and socially compatible management approaches; consequently the concept of agroforestry gained greater acceptance in the industrialized nations during the 1990s. Development of agroforestry systems and their application in countries and regions, such as the United States, China, Australia and New Zealand, and southern Europe, have now clearly demonstrated the range of conditions under which agroforestry can be successfully applied and the myriad of benefits that can be derived (Nair, 2001; Garrett et al., 2000).

THE PROMISE OF AGROFORESTRY

Agroforestry is based on the premise that land-use systems that are structurally and functionally more complex than either crop or tree monocultures result in greater efficiency of resource (nutrients, light, and water) capture and utilization, and greater structural diversity that entails a tighter coupling of nutrient cycles. Above- and below-ground diversity provides more system stability and resilience at the site-level. At the landscape and watershed levels, such systems can provide connectivity with forests and other landscape features to achieve desired ecological services, such as protection of wildlife habitat and water- and soil quality (Ruark et al., 2003).

Agroforestry systems in different parts of the world vary in nature, complexity, and objectives (Nair, 1993). The economic advantage of diversified income is a major motivation for practicing such systems in both tropical and temperate regions. In general, subsistence farming and emphasis on the role of trees in improving soil quality of agricultural lands are characteristic of tropical agroforestry systems. Environmental sustainability is a major driving force for the development and adoption of agroforestry in the industrialized nations (Figure 1), where monocultural production of agriculture and forestry commodities has led to reduced biodiversity and loss of wildlife habitat, increased non-point source pollution of ground water and rivers, and deterioration of family farms. Such problems are the legacy of maximizing production of agricultural products without sufficient knowledge of, or regard for, impacts on future productivity, the environment, and society in general. Today, consequent to the increasing awareness of long-term adverse impacts of current land-use systems, the idea of incorporating the structure and functions of naturally occurring agroecosystems into the design of managed ecosystems is gaining wider acceptance (Lefroy et al.,

FIGURE 1. A (top): A degraded agricultural stream (Washington creek), Ontario, Canada, in 1985. B (bottom): The same Washington creek site in 1990, five years after establishing the riparian tree buffer, during which period the water chemistry along the rehabilitated area improved significantly.

Source: A.M. Gordon and N.V. Thevathasan, University of Guelph, ON, Canada.

1999). This leads to a shift away from separating land uses on discrete parcels to integrating them on a landscape level.

AGROFORESTRY AND FOOD SECURITY

Trees in Support of Food Production

One of the tree-mediated benefits of considerable advantage in the tropics is that trees and other vegetation improve the productivity of the soil beneath them. Research results during the past two decades show that three main tree-mediated processes determine the extent and rate of soil improvement in agroforestry systems. These are: (1) increased N input through biological nitrogen fixation by nitrogen-fixing trees (NFTs), (2) enhanced availability of nutrients resulting from production and decomposition of tree biomass, and (3) greater uptake and utilization of nutrients from deeper layers of soils by deep-rooted trees (Buresh and Tian, 1998; Rao et al., 1998). Furthermore, presence of deep-rooted trees in the system can contribute to improved soil physical conditions and higher soil microbiological activities under agroforestry (Buresh and Tian, 1998). NFTs that are common in the tropics are a particularly valuable resource for soil improvement. Farmlands in many parts of the developing world generally suffer from the continuous depletion of nutrients as farmers harvest without fertilizing adequately or fallowing the land. For example, annual nutrient depletion rates of 22 kg nitrogen, 2.5 kg phosphorus, and 15 kg potassium per hectare of cultivated land during the past 30 years in 37 African countries amount to an annual loss equivalent to the application of $4 billion worth of fertilizers (Sanchez, 2002). Commercial fertilizers cost two to six times as much in Africa as in Europe or Asia. Even at these prices, supplies are problematic due to poorly functioning markets and road infrastructure. One promising way for overcoming this problem is to enable smallholders to use fertilizer tree systems that increase on-farm food production (Sachs, 2005). After years of experimentation with a wide range of soil-fertility replenishment practices, three major types of simple, practical, fertilizer-tree systems have been developed: (1) improved fallows using trees and shrubs, such as sesbania (*Sesbania sesban*) or tephrosia (*Tephrosia vogelii*), (2) mixed intercropping with species such as gliricidia (*Gliricida sepium*), and (3) biomass transfer with species such as wild sunflower (*Tithonia diversifolia*) or gliricidia. These practices can provide 50 to 200 kg N ha^{-1} to the associated cereal crops (Buresh and Tian, 1998), and they tend to be adopted, to a greater extent, by the poorest farm families, which is an unusual experience for agricultural innovations (Garrity, 2004).

The other major avenue of soil improvement with agroforestry is through soil conservation. About 1.9 billion ha, a third of total farmland, in developing nations are estimated to be degraded through erosion, salinity, and fertility depletion (Lal, 2004). Brown (2003, 2004) has vividly portrayed the looming dangers facing humanity from eroding soils and advancing deserts. Drawing from historical examples–from the decline of the Mayan empire (that flourished from the sixth century B.C. to the ninth century A.D.) to the U.S. Dust Bowl of the Great Plains in the 1930s and the Soviet dust bowl in the Virgin Lands in the 1960s, and the current situations in many countries of Africa and Asia–Brown presents an extremely gloomy scenario and calls for action to avert the danger. The potential of agroforestry to reduce the hazards of erosion and desertification as well as to rehabilitate such degraded land and to conserve soil and water has been well recognized (Young, 1997). While this approach is more recent in the temperate zone, research has demonstrated that grass-shrub-tree buffers (agroforestry design) are superior to grass buffers in reducing sediment losses in the temperate zone, too (Schmitt et al., 1999). Direct or supplementary use of trees and shrubs to control soil erosion is now a widespread agroforestry practice in both tropical and temperate regions (Nair, 2001).

Underexploited Trees

Several indigenous agroforestry systems involve a multitude of lesser-known woody species that have come to be known as "multipurpose trees" or "multipurpose trees and shrubs" (MPTs) (Nair, 1993). For example, a large number of fruit-producing trees are integral parts of traditional homestead and other agroforestry farming systems with their characteristic multi-strata canopies, in many developing countries (Figure 2). Table 1 contains such common fruit trees, their ecological and geographical distribution, and the nutritive value of their fruits. The table is, unfortunately, incomplete in respect of "basic" information, such as average yield, life span, without which the trees' contribution to nutritional and food security cannot be convincingly projected. This is but a reflection of the lack of research attention these species have received.

FIGURE 2. Several fruits, many of them from relatively lesser known trees, are an essential part of food security and livelihood of many societies especially in the tropics and subtropics, where such fruit trees are an integral part of traditional multistrata agroforestry systems. The fruits in the picture include durian, *Durio zibethinus*, three large spiny fruits on the right, and rambutan, *Nephelium lappaceum*, a bunch of small spiny fruits at the bottom center, that are included in Table 1.

Another group of underexploited species of immense cultural and economic value is the natural medicinal plants ("medicinals") that are the source of treatment for many diseases and ailments endured by the poor throughout the developing world (Alston and Pardey, 2001; Rao et al., 2004). In Table 2, some examples of commercially valuable medicinal plants that are or can be under cultivation as understory components

TABLE 1. Some underexploited fruit trees grown in tropical and subtropical agroforestry systems.

Species	Distribution		Chemical Composition of Fruit (%)					Energy Value (kJ 100 g^{-1})	High Content of
	Ecological	Geographic	Protein	Ash	Fat	Fiber	Total Carb		
Adansonia digitata	Humid, semiarid	Africa	na	50.0	4.3	8.3	79.4	1480	Ca, Mg, K
Aegle marmelos	Humid-upland	Asia	2.6	-.7	0.4	na	31.8	na	Vitamin C
Aleurites moluccana	Humid, sub-humid	Asia, Africa, Carrib.	19.0	3.0	63.0	na	8.0	na	P, Ca, Fe
Allanblackia floribunda	Sub-humid	Africa	na	na	58.0	na	na	na	
Annona muricata	Humid	Asia, L. Am.	1.0	0.6	na	1.1	14.9	na	
Annona senegalensis	Sub-humid	Africa	na	5.0	8.4	17.8	53.7	1426	Mg, K
Annona squamosa	Sub-humid	Asia, L. Am.	1.6	na	4.0	31.0	23.5	104	Fe
Ardisia pyramidalis	Humid	Asia	13.5	6.1	1.0	37.9	na	na	
Artocarpus heterophyllus	Humid	Asia, L. Am.	1.9	10.0	30.0	11.0	25.4	na	K
Azanza garckeana	Sub-humid	Africa	na	6.3	1.1	45.3	35.2	810	P, Mg, K
Bauhinia thonningii	Humid	Africa	na	2.4	5.9	25.8	63.3	1276	Mg, K
Caryocar brasiliense	Tropics	L. Am.	2.7	10.0	8.0	na	6.7	na	
Caryocar villosum	Tropics	L. Am.	3.0	na	na	14.0	11.0	na	
Chrysophyllum cainito	Sub-humid	Asia, L. Am., Carrib.	na	na	na	na	na	na	Ca
Dacryodes edulis	Humid	Africa, Asia	33.8	12.6	na	27.3	7.6	na	
Durio zibethinus	Upland humid	SE Asia	2.5	na	1.6	na	28.3	na	K
Eugenia stipitata	Upland humid	L. Am., Carrib.	10.8	na	na	6.5	72.0	na	
Flacourtia indica	Humid	Africa	4.2	5.7	3.6	5.7	80.7	1290	P, Mg, K
Garcinia kola	Humid	Africa	7.8	na	na	na	na	na	Vit. C, P, Fe
Genipa americana	Humid	L. Am., Carrib.	na	12.0	na	94.0	13.0	na	Fe
Gnetum africanum	Humid	Africa	20.4	4.7	na	15.2	12.5	na	
Hymenaea courbaril	Sub-humid	L. Am., Asia	5.9	2.0	2.2	13.4	75.3	na	P
Inga edulis	Sub-humid	L. Am., Carrib.	11.7	1.7	0.8	2.8	39.5	na	
Irvinga gabonensis	Sub-humid	Africa	7.4	2.5	51.3	0.9	26.0	na	
Jeasenia bataua	Humid	L. Am., Carrib.	0.6	1.7	na	49.5	51.5	na	
Lansium domesticum	Upland	Asia	0.8	na	na	2.3	9.5	na	
Morus nigra	Sub-humid	L. Am., Asia	1.5	80.0	0.5	1.4	8.3	na	
Nephelium lappaceum	Sub-humid	SE Asia	0.5	na	na	0.2	16.0	na	K
Parinari curatellifolia	Sub-humid	Africa	3.0	1.8	1.5	5.5	88.2	1517	Vit. C
Parkia biglobosa	Humid	Africa, Carrib.	40.0	na	na	na	60.0	na	
Poraqueiba sericea	Sub-humid	L. Am.	3.5	1.0	na	17.0	na	na	Vit. C
Pouteria campechiana	Upland	L. Am.	1.7	na	0.1	0.1	36.7	na	Vit. C
Sclerocarya birrea	Sub-humid	Africa	31.0	na	61.0	na	3.7	na	

TABLE 1 (continued)

| Species | Distribution | | Chemical Composition of Fruit (%) | | | | | Energy Value | High |
	Ecological	Geographic	Protein	Ash	Fat	Fiber	Total Carb	(kJ 100 g^{-1})	Content of
Spondias mombin	Sub-humid	L Am., Carrib.	1.4	31.4	0.6	1.2	10.0	na	P, Mg, K
Strychnos innocua	Humid	Africa	11.5	3.7	6.0	17.9	61.0	1390	P, K
Strychnos spinosa	Humid	Africa	5.4	4.1	31.2	17.6	42.1	1923	Mg, K
Syzigium guineense	Sub-humid	Africa	10.1	7.1	4.0	30.3	48.5	1093	P, Mg, K
Tamarindus indica	Sub-humid	Pan-tropical	4.1	3.4	1.6	5.9	85.0	1490	P, Mg, K
Trichilia emetica	Humid	Africa	17.0	4.5	22.9	8.1	47.5	1897	P, Mg, K
Uapaka kirkiana	Sub-humid	Africa	1.8	2.2	1.1	8.4	86.5	1456	Vit. C, Mg
Vangueria infausta	Semiarid	Africa	5.7	3.4	2.6	10.2	78.1	1445	Mg, K
Vitellaria paradoxa	Semiarid	Africa	na	na	27.3	na	na	na	
Vitex doniana	Sub-humid	Africa	2.6	4.8	0.7	5.2	86.7	1459	P, K
Ximenia caffra	Sub-humid	Africa	7.6	11.0	5.2	2.3	78.8	1506	P, K
Ziziphus mauritiana	Semiarid	Pan-tropical	4.1	10.1	9.5	3.4	73.0	1588	P, K

Notes:

1. These data are compiled from various sources, such as ICRAF (World Agroforestry Centre) (http://www.worldagroforestry.org/Sites/TreeDBS/TreeDatabases.asp), Purdue University (http://www.hort.purdue.edu/newcrop), FAO, Leakey (1999), Saka and Msonthi (1994), and various on-line sources. The authenticity and accuracy of the data may, however, be questionable, considering that the various sources may have relied on previous data sources and thus most of the information could be based on FAO compilations of the 1980s.

2. The species were selected based on (1) FAO publication (Food and Fruit-Bearing Forest Species) series, (2) World Agroforestry Congress Abstracts, (3) World Agroforestry Centre Agroforestree Database, (4) Forestry Compendium (2005), and (5) journal articles. Geographical distribution of each species was determined from the Forestry Compendium and from the World Agroforestry Centre Agroforestree Database.

3. For all species, the primary edible parts are their fruits (or parts thereof such as fruit pulp, seed, or other parts). In some cases, leaves are also edible (e.g., *Adansonia digitata, Aegle marmelos, Azanza garckeana, Bauhinia thonningii, Gnetum africanum, Parkia biglobosa, Tamarindus indica*).

4. "na" means data not available.

5. Chemical composition is based on fruit dry weight; conversion was made for some values to arrive at a common unit for comparison.

Major citations (specific to this table):

- ICRAF (World Agroforestry Centre): (http://www.worldagroforestry.org/Sites/TreeDBS/TreeDatabases.asp)
- Purdue University (http://www.hort.purdue.edu/newcrop)
- FAO (www.fao.org)
- Leakey R.R.B. 1999. Potential for novel food products from Agroforestry trees: A review. Food Chemistry 66: 1-14
- Saka J.D. and J.D. Msonthi. 1994. Nutritional value of edible fruits of indigenous wild trees in Malawi. Forest Ecology and Management 64: 245-248
- Other online sources
  http://www.answers.com/topic/brazil-nut
  http://www.bawarchi.com/health/sitaphal.html
  http://www.montosogardens.com/durio_zibethinus.htm
  http://www.soton.ac.uk/~icuc/dacrbib/dacr-u-f1.htm

TABLE 2. Examples of commercially valuable medicinal plants under cultivation or that can be produced as understory components in forests and tree plantations.

Latin name	Common name	Plant type	Parts used	Medicinal use	Location
Amomum subulatum	Large cardamom	Perennial herb	Seeds	Stimulant, indigestion, vomiting, rectal diseases	Sub-Himalayan range, Nepal, Bhutan
Amomum villosum	'Saren'	Perennial herb	Seeds	Gastric and digestive disorders	China
Caulophyllum thalictroides	Blue cohosh	Perennial herb	Roots	Gynecological problems, bronchitis	North America
Cephaelis ipecacuanha	Raicilla, Ipecac	Shrub	Roots	Whooping cough, bronchial asthma, amoebic dysentery	Brazil, India, Bangladesh, Indonesia
Cimicifuga racemosa	Black cohosh	Perennial herb	Roots	Menses related problems	North America
Chlorophytum borivilianum	'Safed musli'	Annual herb	Tubers	Male impotency, general weakness	India
Dioscorea deltoidea	Himalayan yam	Vine	Tubers	Source of saponins and steroids	India, Pakistan
Echinacea purpurea	Coneflower	Perennial herb	Roots, rhizomes	Enhancing immune system	North America
Hydrastis canadensis	Goldenseal	Perennial herb	Rhizomes	Tonic	North America
Panax ginseng	Ginseng	Herb	Roots	Tonic	China, Korea, Japan
Panax quinquefolium	American ginseng	Perennial herb	Root	Tonic	North America
Rauvolfia serpentina	Rauvolfia	Shrub	Roots	Hypertension and certain forms of insanity	Sub-montane zone, India
Serenoa repens	Saw palmetto	Shrubby-palm	Fruits	Swelling of prostrate gland	Southeastern USA

Adapted from Rao et al. (2004).

in forests and tree plantations, are given. Additionally, a large number of species, often grown in backyards or home gardens, or in natural forests, are used for medicinal purposes in several indigenous communities. In Africa, for example, more than 80% of the population depends on medicinal plants to meet their medical needs, and about two-thirds of the species from which such medicines are derived are trees (Rao et al., 2004). While the majority of these tree products are obtained by extraction from natural forests, some 'well-known' agroforestry tree species grown on farms for other uses (such as fodder, food, or fuel-wood) are also used for their medicinal values. Examples include *Acacia nilotica*, the gum of which is used in India and Africa for treating diarrhea, dysentery, diabetes, sore throat, and its bark used to arrest external bleeding; *Azadirachta indica*, the neem tree, used throughout Asia and Africa against digestive disorders, malaria, fever, hemorrhoids, hepatitis, measles, syphilis, boils, burns, snakebite, and rheumatism); *Parkia biglobosa* (*neré* or the locust bean tree) used in Africa for treatment of piles, malaria, stomach disorders, and jaundice; and *Tamarindus indica*, the tamarind tree, used in India and Africa as a refrigerant, laxative, and carminative as well as in febrile diseases and bilious disorders (Rao et al., 2004). There is also increasing interest in natural medicines in the developed world, creating new or expanded markets for these products. This puts further extraction pressure on the natural forests. Many of the medicinal tree species are already overexploited. Some species are so depleted that their gene pools are greatly eroded (e.g., *Prunus africana*), and some are in danger of extinction (Alston and Pardey, 2001; Simons and Leakey, 2004).

Significant efforts are needed to domesticate new and underutilized tree species, to intensify their cultivation on smallholder farms, and to develop market infrastructures. Impact studies conducted on return to investments in research on trees and tree crops report internal rates of returns averaging 88%, which compare favorably with the returns on research on field crops averaging 74% (Alston and Pardey, 2002). Research on issues such as tree enterprise development and tree-product marketing enhancement is practically unheard of. Rural areas in the tropics typically have markets with high risks and high transaction costs. This makes production diversification a preference of small-scale cash-limited farmers (Garrity, 2004). Under such conditions, integrated agroforestry systems are a suitable pathway toward improved livelihoods. Public-private partnerships that have been successful in the development of tree commodities, such as cacao (*Theobroma cacao* L.), could serve as a model for future development of agroforestry tree crops (Shapiro and Rosenquist, 2004).

ENVIRONMENTAL SERVICES OF AGROFORESTRY

Biological Diversity in Working Landscapes

Many landscapes do not have adequate forested habitat to satisfy the requirements of some species of plants and animals, and available forest reserve areas may be too small to contain the habitat requirements of all species (Ruark et al., 2003). Agroforestry can provide ways of augmenting the supply of forest habitat and providing greater landscape connectivity. Where croplands occupy most of the landscape, riparian forest buffers and field shelterbelts can be essential for maintaining plant and animal biodiversity, especially under a changing climate scenario. A comprehensive assessment of shelterbelt agroforestry systems in the northern Great Plains of the USA has clearly demonstrated their importance for breeding bird species richness and community composition at both the farm and landscape levels (Pierce et al., 2001). Agroforestry adds plant and animal biodiversity to landscapes that might otherwise contain only monocultures of agricultural crops (Schroth et al., 2004). Growing commercial crops, such as coffee (*Coffea* sp.) and cacao, known as shaded-perennial systems in agroforestry literature (Nair, 1993), is projected to start a trend of combining environmental research with consumer products, which could then have a large impact on global conservation.[4]

The well being of the land is directly tied to the well being of its inhabitants. Providing rural people and poor farmers with the opportunity to earn sustainable, stable livelihoods will help conserve the planet's biodiversity (Schroth et al., 20004). As much as 90 percent of the biodiversity resources in the tropics are located in human-dominated or working landscapes (McNeely, 2004). Agroforestry impinges on biodiversity in working landscapes in at least three ways. First, the intensification of agroforestry systems can reduce exploitation of nearby or even distant protected areas. Second, the expansion of agroforestry systems into traditional farmlands can increase biodiversity in working landscapes. Third, agroforestry development may increase the species and within-species diversity of trees in farming systems (McNeely, 2004).

A new paradigm is emerging that integrates protected areas into their broader landscapes of human use and biodiversity conservation, particularly in agricultural areas that now constitute the principal land use in most of the developing world (Garrity, 2004). The issue of how best to achieve a balance between production and biodiversity conservation has become the basis for the concept of *ecoagriculture*, which refers to land-use systems managed for both agricultural production and wild biodiversity conservation (McNeely and Scherr, 2003). Agroforestry is uniquely suited to provide ecoagriculture solutions (McNeely, 2004), but more needs to be understood before widespread application of the solutions can be recommended.

Carbon Storage

Agroforestry systems can sequester substantial quantities of carbon (Dixon, 1995; Lal et al., 2004). Trees and shrubs planted in shelterbelts store carbon in their shoots and roots, while protecting soils and crops (Young, 1997) and providing biodiversity and habitat for wildlife (Schroth et al., 2004). Through either deposition of wind-blown soils or interception of surface runoff sediments, many of the linear-based agroforestry practices, such as shelterbelts and riparian buffers, trap significant amounts of carbon-rich topsoil that would otherwise be lost from these systems (Pallardy et al., 2003). Riparian forest buffers are natural carbon sinks (Montagnini and Nair, 2004). When suitable trees and shrubs are grown in these moist environments, they filter out contaminants from adjacent agricultural or community activities (Schultz et al., 2004).

In temperate regions, agroforestry practices have been estimated to have the potential to store C in the range of 15 to 198 Mg C ha^{-1} (mode: 34 Mg C ha^{-1}) (Pandey, 2002). The C sequestration potential of agriculture in the United States through the application of agroforestry practices by 2025 is estimated at 90.3 Tg C y^{-1} (Nair and Nair, 2003). In the tropics, agroforestry systems are estimated to have helped regain 35% of the original C stock of the cleared forest, compared to only 12% by croplands and pastures (Palm, 2004). A projection of carbon stocks for smallholder agroforestry systems indicates C sequestration rates ranging from 1.5 to 3.5 Mg C ha^{-1} yr^{-1} and a tripling of C stocks in a 20-year period to 70 Mg C ha^{-1} (Palm, 2004). Global deforestation, which is estimated to occur at the rate of 17 million ha yr^{-1}, is expected to cause the emission of 1.6 Pg C ha^{-1} yr^{-1}. Assuming that one hectare of agroforestry could save five hectares from deforestation, carbon emission caused by deforestation could be reduced substantially by establishing agroforestry systems (Lal et al., 2004; Palm et al., 2004).

Water Quality and Environmental Amelioration

The effectiveness of some agroforestry practices, such as riparian buffers in reducing non-point source pollution, and thereby improving water quality, is well documented (Udawata et al., 2002; Schultz et al., 2004). The deeper and more extensive tree roots will invariably be able to take up more nutrients (especially N) from the soil compared to crops with shallower root systems—the so-called "safety-net" effect that has been affirmed in various agroforestry situations (Buresh and Tian, 1998; Jose et al., 2004). Consequently, nutrient-leaching rates from soils under agroforestry systems where trees are a major component can be lower than those from treeless systems (Nair and Graetz, 2004). The water-quality enhancement resulting from the reduction of nutrient loading could be a substantial environmental benefit of agroforestry in heavily fertilized agricultural landscapes.

SCALING UP THE BENEFITS OF AGROFORESTRY

Achieving the full promise of agroforestry requires fundamental understanding of how and why farmers make long-term land-use decisions and applying this knowledge to the design, development, and 'marketing' of agroforestry innovations (Franzel et al., 2004). Following this realization, social science research–adoption research in particular–has received substantial attention in the agroforestry research agenda during the past decade (Mercer, 2004). An area that needs special attention, however, is policy research. It is crucial to have enabling polices in place to absorb the results of biophysical research and incorporate them into government actions. For example, if the global environmental services provided by agroforestry, such as carbon sequestration and water-quality improvement, are not internalized to the benefit of land users, they will have little incentive to adopt agroforestry at a socially desirable level. Recent advancements in environmental economics, especially valuation methodologies (Alavalapati and Mercer 2004), help quantify the potential demand for and supply of environmental services more accurately. For example, the impact of a tax on phosphorus runoff and a payment for carbon sequestration on the profitability of traditional cattle ranching and silvopasture, a form of agroforestry, was investigated in southern Florida. For all pollution tax rates with carbon payments of greater than $1 per ton ($10^6$ gram), land values under silvopasture are greater than those with traditional ranching. On average, a price premium of $0.33/kg ($0.15/lb) of beef, or a direct payment of $23.00/ha ($9.32/acre) annually, was found to influence ranchers to adopt silvopasture practices (Shrestha and Alavalapati, 2004). These estimates of incentive levels are much lower compared with previous studies on farmers' willingness to participate in conservation programs in the United States, suggesting the possibilities for wider acceptance of agroforestry practices. Policy makers can use this information as a basis to formulate incentive or tax policies to the benefit of the economy and the environment.

AGROFORESTRY: A PANACEA?

In promoting agroforestry, more harm than good has been done in many cases by making exaggerated claims about its role and potential as a land-management approach. Agroforestry is certainly not a "cure-all" for all the problems of land management. It is not a substitute for agriculture or forestry, as we know them, nor is it for all situations. But it certainly has a role–more as a complementary and supportive than principal one–in some situations, especially under resource-poor and limited and low-input conditions.

Impatience arising from the long-term nature of agroforestry research, often fueled by the short-term nature of research funding and, sometimes, unrealistic expectations and deadlines placed by donor agencies, forces researchers to come up with short-term claims that are not borne out by good science. But, practices whose underlying scientific principles have not been substantiated through rigorous scientific procedures will not stand the test of time. While it is creditable that considerable progress has been achieved during the past 25 years in transferring the age-old agroforestry practices into a science-based activity, several knowledge gaps exist even in areas that have received research attention in the past. Then there are several potentially promising areas that have not been explored. For example, practically no research has been done on the several tree products that are known to be an essential part of the traditional diet, health care, and livelihood of people in many societies. Substantial efforts are needed to domesticate indigenous fruit and medicinal trees and promote their cultivation on farms. Research partnerships between agroforestry and the medical and nutritional sciences and the food products industry will be crucial to ensure that the key tree species for such uses are developed for farm cultivation. In our obsession with "grain crops" in modern agriculture, we have ignored them. Ecological advantages of agroforestry systems involving such trees (e.g., enhanced nutrient cycling, reduced dependence on external nutrient inputs because of reduced "export" through harvested products, enhanced biodiversity, and so on) have only been hypothesized but not adequately documented. The exploitation of these species and the agroforestry practices involving their use has wide implications in food security and environmental protection, as well as conservation and use of genetic resources. A new "tree crops revolution" is needed that broadens the array of tree products that are produced, processed and delivered by developing countries to regional and global markets. These products could be a vehicle providing comparative advantage to poor land-locked countries, and countries that are not otherwise integrated into the global economy. For promoting these tree products, a new research-and-development strategy that reduces dependency on primary agricultural commodities is needed; development institutions must incorporate new skills in domestication of indigenous species (Simons and Leakey, 2004) and the processing/storage of their products, and in market analysis and market linkages.

CONCLUSIONS

Some of the major threats facing the world today, such as food and nutritional security, eroding soils, and expanding deserts, have not attracted deserving attention of many national governments and international development agencies. The time has arrived for utilizing the benefits of the remarriage of crops and trees in addressing some of these "mega threats." Too often, we treat agriculture and forestry separately, yet these two sectors are often interwoven on the landscape and share many common goals. The scope of the successful Green Revolution in the tropics needs to be broadened to include trees as an essential part of farming systems. Agroforestry systems are considered to have the potential to help alleviate hunger and poverty and provide food security to 1.8 billion people in the developing countries. If we are to meet society's needs and aspirations for forest-derived goods and services, we must find ways of augmenting traditional forestry by gleaning some portion of these benefits from agricultural lands where agroforestry can be practiced. Indeed, in many places the only opportunity to provide increased forest-based benefits, such as wildlife habitat or forested riparian systems, is through the increased use of agroforestry on agricultural lands. On the other hand, in many forest-based ecosystems, agroforestry principles are being employed to derive benefits, such as non-timber forest products. In developed countries, such as the United States, current interest in ecosystem management strongly suggests that there is a need to embrace and apply agroforestry principles to help mitigate non-point source pollution and other environmental problems and better meet the current and future needs for the products and services of the land.

Twenty-five years ago, agroforestry began to attract the attention of the international development and scientific community, primarily as a means for sustaining agricultural production in marginal lands and remote areas of the tropics that were not benefited by the Green Revolution. Thanks to the modest research input, today agroforestry has been recognized to have the potential to offer much more toward ensuring not only food security in poor countries, but also environmental integrity in poor and rich nations alike. Agroforestry has come of age. Enhanced investments in research and development are needed to capitalize on the enormous potentials offered by agroforestry.

NOTES

1. But, the era of rapidly growing fertilizer use seems to be over. For example, US fertilizer use is essentially the same today as it was in the early 1980s, between 17 million and 21 million tons a year, and the situation is the same in Western Europe and Japan. Among the three major grain-producing countries, China is the leading fertilizer user, with the USA way behind and India set to overtake it. In many countries that already have effectively removed nutrient constraints on crop yields, applying more fertilizer has little effect on yields. The adverse effects of excessive fertilizer application are also increasingly being recognized.

2. However, as with fertilizer use, growth in irrigation has declined dramatically worldwide during the past decade.

3. The Congress was hosted by the University of Florida and co-sponsored by more than 25 public and private institutions from four continents (http://conference.ifas.ufl.edu/wca). Inaugurating the Congress, Norman Borlaug said, "Peace and stability in the world cannot be built on human misery in the Third World. Agroforestry would continue to have an important role in alleviating both poverty and environmental degradation in low-income countries." In his plenary address to the Congress, M. S. Swaminathan (www.mssrf.org) urged the world community to consider agroforestry as a means for "bio-happiness."

4. *Science* 305: 941 (2004).

REFERENCES

Alavalapati, J.R.R., and D.E. Mercer (ed.). 2004. Valuing agroforestry systems: Methods and applications. (Advances in Agroforestry # 2). Kluwer, Dordrecht, The Netherlands. 314 p.

Alston, J.M., and P.G. Pardey. 2001. Attribution and other problems in assessing the returns to agricultural R & D. Agric. Econ. 25: 141-162.

Bene, J.G., H.W. Beall, and A. Côte. 1977. Trees, food and people. International Development Research Centre, Ottawa, Canada. 52 p.

Brown, L.R. 2003. Plan B: Rescuing a planet under stress and a civilization in trouble. W.W. Norton, New York. 285 p.

Brown, L.R. 2004. Outgrowing the earth: The food security challenge in an age of falling water tables and rising temperatures. W.W. Norton, New York. 239 p.

Buresh, R.J., and G. Tian. 1998. Soil improvement by trees in sub-Saharan Africa. Agrofor. Syst. 38: 51-76.

Dixon, R.K. 1995. Agroforestry systems: Sources or sinks of greenhouse gases? Agrofor. Syst. 31: 99-116.

Evenson, R.E., and D. Gollin. 2003. Assessing the impact of the Green Revolution, 1960 to 2000. Science 300: 758-762.

Franzel, S., G.L. Denning, J.P.B. Lilleso, and A.R. Mercardo, Jr. 2004. Scaling up the impact of agroforestry: Lessons from three sites in Africa and Asia. Agrofor. Syst. 61: 329-344.

Garrett, H.E., W.J. Rietveld, and R.F. Fisher (eds.). 2000. North American agroforestry: An integrated science and practice. Am Soc. Agron. Madison, WI. 402 p.

Garrity, D.P. 2004. Agroforestry and the achievement of the Millennium Development Goals. Agrofor. Syst. 61: 5-17.

Jose, S., A.R. Gillespie, and S.G. Pallardy. 2004. Interspecific interactions in temperate agroforestry. Agrofor. Syst. 61: 237-255.

Lal, R. 2004. Soil carbon sequestration impacts on global climate change and food security. Science (Washington, DC) 304: 1623-1627.

Lefroy, E.C., R.J. Hobbs, M.H. O'Connor, and J.S. Pate (eds.). 1999. Agriculture as a mimic of natural ecosystems. Kluwer, Dordrecht, The Netherlands. 492 p.

McNeely, J.A. 2004. Nature vs. nurture: Managing relationships between forests, agroforestry and wild biodiversity. Agrofor. Syst. 61: 155-165.

McNeely, J.A., and S.J. Scherr. 2003. Ecoagriculture: Strategies to feed the world and save wild biodiversity. Island Press, Washington, DC. 323 p.

Mercer, D.E. 2004. Adoption of agroforestry innovations in the tropics. Agrofor. Syst. 61: 311-328.

Montagnini, F., and P.K.R. Nair. 2004. Carbon sequestration: An under-exploited environmental benefit of agroforestry systems. Agrofor. Syst. 61: 281-298.

Nair, P.K.R. 1993. An introduction to agroforestry. Kluwer, Dordrecht, The Netherlands. 499 p.

Nair, P.K.R. 2001. Agroforestry. In Our fragile world: Challenges and opportunities for sustainable development, forerunner to the encyclopedia of life support systems, M. Tolba (ed.), Chapter 1.25, pp. 375-393, vol. I. UNESCO, Paris, France, and EOLSS, Oxford, UK.

Nair, P.K.R., M.E. Bennister, and S.C. Allen. 2005. Agroforestry today: An analysis of the 750 papers presented to the 1st World Congress of Agroforestry, 2004. J. For. 103: 417-421.

Nair, P.K.R., and V.D. Nair. 2003. Carbon storage in North American agroforestry systems. In J. Kimble, L.S. Heath, R.A. Birdsey, and R. Lal (ed.). The potential of U.S. forest soils to sequester carbon and mitigate the greenhouse effect, pp. 333-346. CRC Press, Boca Raton, FL.

Nair, V.D., and D.A. Graetz. 2004. Agroforestry as an approach to minimizing nutrient loss from heavily fertilized soils: The Florida experience. Agrofor. Syst. 61: 269-279.

Pallardy, S.G., D.E. Gibbons, and J.L. Rhoads. 2003. Biomass production by two-year-old poplar clones on floodplain sites in the lower Midwest, USA. Agrofor. Syst. 59: 21-26.

Palm, C.A., T. Tomich, M. Van Noordwijk, S. Vosti, J. Gockowski, J. Alegre, and L. Verchot. 2004. Mitigating GHG emissions in the humid tropics: Case studies from the Alternatives to Slash-and-Burn Program (ASB). Environ. Dev. Sust. 6: 145-162.

Pandey, D.N. 2002. Carbon sequestration in agroforestry systems. Clim. Policy 2: 367-377.

Pierce, R.A. II, D.T. Farrand, and W.B. Kurtz. 2001. Projecting the bird community response resulting from the adoption of shelterbelt agroforestry practices in Eastern Nebraska. Agrofor. Syst. 53: 333-350.

Rao, M.R., P.K.R. Nair, and C.K. Ong. 1998. Biophysical interactions in tropical agroforestry systems. Agrofor. Syst. 38: 3-50.

Rao, M.R., M.C. Palada, and B.R. Becker. 2004. Medicinal and aromatic plants in agroforestry systems. Agrofor. Syst. 61: 155-165.

Ruark, G.A., M.M. Schoeneberger, and P.K.R. Nair. 2003. Roles for agroforestry in helping to achieve sustainable forest management. U.N. Forum on Forests (UNFF) Intersessional Experts Meeting, 24-30 March 2003, Wellington, New Zealand [Online]. Available at http://www.maf.govt.nz/mafnet/unff-planted-forestry-meeting/conference-papers/roles-for-agroforestry.htm (verified 14 September 2005).

Sachs, J.D. 2005. The end of poverty: Economic possibilities for our time. Penguin, New York. 320 p.

Sanchez, P.A. 2002. Soil fertility and hunger in Africa. Science 295: 2019-2020.

Schmitt, T.J., M.G. Dosskey, and K.D. Hoagland. 1999. Filter strip performance and processes for different vegetation, widths and contaminants. J. Environ. Qual. 28: 1479-1489.

Schroth, G., A.B. da Fonseca, C.A. Harvey, C. Gascon, H.L. Vasconcelos, and N. Izac (eds.). 2004. Agroforestry and biodiversity conservation in tropical landscapes. Island Press, Washington, DC. 523 p.

Schultz, R.C., T.M. Isenhart, W.W. Simpkins, and P.J. Colletti. 2004. Riparian forest buffers in agroecosystems: Lessons learned from the Bear Creek Watershed, central Iowa, USA. Agrofor. Syst. 61: 35-50.

Shapiro, H.-Y., and E.M. Rosenquist. 2004. Public/private partnership in agroforestry: The example of working together to improve cocoa sustainability. Agrofor. Syst. 61: 453-462.

Shrestha, R.K., and J.R.R. Alavalapati. 2004. Valuing environmental benefits of silvopasture practice: A case study of the Lake Okeechobee Watershed in Florida. Ecol. Econ. 49: 349-359.

Simons, A.J., and R.R.B. Leakey. 2004. Tree domestication in tropical agroforestry. Agrofor. Syst. 61: 167-181.

Spurgeon, D. 1979. Agroforestry: new hope for subsistence farmers. Nature 280: 533-534.

Steppler, H.A., and P.K.R. Nair (eds.). 1987. Agroforestry: A decade of development. ICRAF (World Agroforestry Centre), Nairobi, Kenya. 335 p.

Udawatta, R.P., J.J. Krstansky, G.S. Henderson, and H.E. Garrett. 2002. Agroforestry practices, runoff, and nutrient loss: A paired watershed comparison. J. Environ. Qual. 31: 1214-1225.

Wackernagel, M., N.B. Schulz, D. Deumling, A.C. Linares, M. Jenkins, V. Kapos, C. Monfreda, J. Loh, N. Myers, R. Norgaard, and J. Randers. 2002. Tracking the ecological overshoot of human economy. Proc. National Acad. Sci. USA, 99 (14): 9266-9271.

Young, A. 1997. Agroforestry for soil conservation. CABI, Wallingford, UK. 276 p.

doi:10.1300/J411v19n01_02

Managing Soils for Food Security
and Climate Change

R. Lal

SUMMARY. Four principal issues facing developing countries, such as
India, are: (i) meeting food demand of the growing population, (ii) reducing risks of soil and ecosystem degradation, (iii) minimizing risks of
eutrophication and contamination of natural waters, and (iv) decreasing
net emissions of CO_2 and other greenhouse gases. A viable solution lies
in a paradigm shift of not taking soils for granted. Soils must be improved, restored and used, rather than depleted, degraded and abused.
Basic principles of soil management are: (i) creating a positive nutrient
balance in agro-ecosystems, (ii) using crop residue, manure and other
bio-solids as components of integrated nutrient management of enhancing nutrient-use efficiency, (iii) improving water-use efficiency,
and (iv) reducing or eliminating plowing. Adoption of these measures
requires a radical shift in the scientific, social, ethnic and cultural fabric
of a society through developing alternatives to: (i) using dung as cooking
fuel, (ii) crop residue as fodder, (iii) top soil for brick making, (iv) plowing for weed control, and (v) flood irrigation for improving soil-water
regime. These practices, used for thousands of years, may have been sustainable when population was low. With large and increasing population, these extractive practices degrade natural resources, pollute the
environment and change the climate. Establishing bio-fuel plantations

R. Lal is affiliated with the Carbon Management and Sequestration Center, The
Ohio State University, Columbus, OH 43210 USA (E-mail: lal.1@osu.edu).

[Haworth co-indexing entry note]: "Managing Soils for Food Security and Climate Change." Lal, R.
Co-published simultaneously in *Journal of Crop Improvement* (Haworth Food & Agricultural Products Press,
an imprint of The Haworth Press, Inc.) Vol. 19, No. 1/2 (#37/38), 2007, pp. 49-71; and: *Agricultural and Environmental Sustainability: Considerations for the Future* (ed: Manjit S. Kang) Haworth Food & Agricultural
Products Press, an imprint of The Haworth Press, Inc., 2007, pp. 49-71. Single or multiple copies of this article are available for a fee from The Haworth Document Delivery Service [1-800-HAWORTH, 9:00 a.m. -
5:00 p.m. (EST). E-mail address: docdelivery@haworthpress.com].

on the village common land and degraded soils, using forage-based crop rotations, making bricks from fly ash and other by-products, adopting no-till farming and using drip or sub-irrigation are high priorities that require serious and immediate attention of policy makers, land managers, farmers and the public at large. While improving agronomic production and advancing food security, soil applications of dung, crop residue and other bio-solids have strong regional and global impacts on the carbon cycle, hydrologic balance, climate and drought, ecosystem restoration and quality of surface and ground waters. doi:10.1300/J411v19n01_03 *[Article copies available for a fee from The Haworth Document Delivery Service: 1-800-HAWORTH. E-mail address: <docdelivery@haworthpress.com> Website: <http://www.HaworthPress.com> © 2007 by The Haworth Press, Inc. All rights reserved.]*

KEYWORDS. Soil carbon sequestration, sustainable agriculture, global warming, conservation tillage, integrated nutrient management, mulch farming

INTRODUCTION

Principal global issues of the 21st century are: (1) population of 6.5 billion and increasing at the rate of 1.14% yr^{-1}, (2) per capita arable land area of 0.22 ha and decreasing to < 0.07 ha for 30 countries by 2025, (3) soil degradation of 2 billion ha (Bha) and increasing at the rate of 5-10 million hectares (Mha) yr^{-1}, (4) renewable fresh water supply of < 1000 m^3 for 30 countries and increasing to 58 countries by 2050, (5) atmospheric CO_2 concentration of 378 ppm and increasing at 0.46% yr^{-1}, (6) energy use of 435 quads (10^{15} BTU) yr^{-1} and increasing at the rate of 2.2% yr^{-1}, and (7) per capita grain consumption of 300 Kg yr^{-1} and decreasing. These global issues cut across national/political borders, because people and nature are inextricably linked, irrespective of political boundaries. Therefore, addressing these issues warrants a coherent and coordinated effort. Further, soil degradation is a common thread that transcends all seven issues. Therefore, an objective assessment of the processes, causes and factors affecting soil degradation may be necessary to identify viable technological interventions of addressing these issues.

The term soil degradation implies decline in soil quality and its ability to perform ecosystem services. Soil degradation is a biophysical process driven by social, economic, political, cultural and ethnic forces.

Important among biophysical factors are climate, land use, land cover, and soil type. All other factors remaining the same, soil degradation risks are more severe in hot and dry than in cool and humid climates. Climatic parameters with strong impact on soil degradation include the amount and seasonal distribution of precipitation, mean annual temperature, and wind velocity and direction. Land use and land cover moderate the effects of climatic parameters. A dense and continuous canopy cover protects soils surface against the degradative effects of climatic parameters, especially that of erosivity (kinetic energy of raindrops and wind). In general, soil is in a dynamic equilibrium with processes under a natural ecosystem (Figure 1). Conversion of natural to managed ecosystems disturbs this equilibrium and sets in motion soil degradative

FIGURE 1. Biophysical, economic and social factors affecting soil degradation.

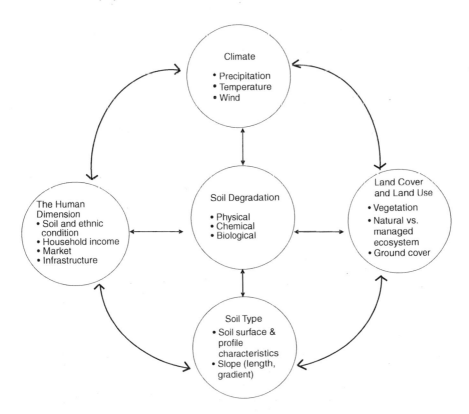

processes (e.g., erosion, disruption in elemental cycling, depletion of soil organic carbon or SOC). Soil type and antecedent properties, both of the surface layer and the profile, determine its susceptibility to degradation processes. For example, soil's susceptibility to crusting, compaction and erosion depend on the degree and strength of aggregates. Susceptibility to erosion also depends on terrain and the physiographic characteristics, especially slope gradient, length and aspect.

The human dimensions, as important as the biophysical factors, need to be addressed especially in terms of the interactive effects (Figure 1). Important among economic factors are the household income that determines farmers' ability to purchase off-farm inputs and invest in soil conservation and restorative measures, access to market, farm size, etc. Social factors relevant to soil degradation are land tenure, gender and equity issues. Political stability impacts and is impacted by soil degradation. Indeed, there is a close relationship between soil degradation and political stability. Regions and nations with severe problems of soil degradation are also politically unstable (Diamond, 2005). Numerous ancient civilizations (e.g., Maya, Mesopotamia, Indus) vanished because they did not pay close attention to the soil that supported them.

The objective of this report is to describe soil management options to address two of the seven global issues: climate change and global food security. This manuscript focuses on the biophysical factors, with specific reference to soil management in the tropics and sub-tropics.

SOIL DEGRADATION AND ECOSYSTEM SERVICES

The schematics in Figure 2 outline adverse impacts of soil degradation on ecosystem services. Soil degradation impacts climate by disrupting the global C cycle and altering the energy balance. Soil degradation disrupts aggregates and exposes the organic matter, hitherto encapsulated and protected within aggregates, to microbial processes. Aggregate disruption usually enhances emission of CO_2. Degradative processes also reduce soil's capacity to oxidize CH_4. Further, degraded soils are prone to denitrification leading to N_2O emission. Therefore, degraded soils may have higher flux of greenhouse gases (GHGs) per unit of net primary productivity (NPP) than undegraded soils. Microclimate of degraded soils is also altered by change in albedo. Depletion of the SOC pool in degraded soil alters its hue and chroma through reduction in humus and change in water retention capacity of the surface layer (Figure 2).

FIGURE 2. Effects of soil degradation on ecosystem services.

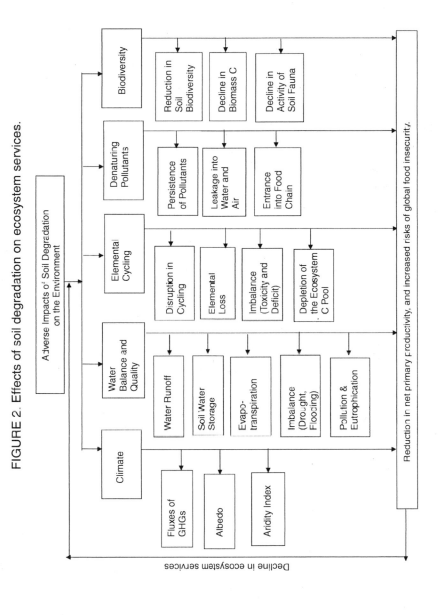

Soil degradation also affects the hydrologic balance. Runoff losses are more from degraded than undegraded soils, and change in energy balance increases evaporation losses. Therefore, soil water storage is lower in degraded than undegraded soils. Increase in water runoff exacerbates transport of dissolved and suspended loads in surface waters causing eutrophication and pollution. Soil's capacity to absorb and denature pollutants (e.g., industrial compounds) is also low on degraded soils because of reduction in faunal activity, and decline in clay and humus concentrations. Leaching of pollutants into the ground water is more in degraded than undegraded soils. One of the strong adverse impacts of soil degradation is on biodiversity. These and numerous other ecosystem services are strongly impacted by soil degradation. An important adverse impact of soil degradation is reduction in biomass production and the NPP, with a strong ramification on regional and global food security (Scherr and Yadav, 1999) (Figure 2). Soil degradation is a notable ramification of human domination of Earth's ecosystems (Vitousek et al., 1997). Thus, restoring value to the world's degraded land (Dailey, 1995) is an ecological and anthropological necessity.

FOOD DEMAND AND THE WORLD POPULATION

Settled agriculture began about 10,000 years ago, when the world population was < 10 million. The world population was 200 to 400 million about 1 A.D., and increased to 1 billion by 1850. A drastic increase in population occurred during the 20th century. The population increased to 2 billion in 1930, 3 billion in 1960, 4 billion in 1975, 5 billion in 1987, 6 billion in 1998, 6.5 billion in 2006 and is increasing at the rate of about 1.3% yr^{-1}. The world population is expected to reach about 7 billion by 2010, 8 billion by 2025 and stabilize at about 10 billion by 2100 (Cohen, 2003; Rees, 2004). During the last 10,000 years, the world population has doubled at least 10 times. However, it will never double again, and will stabilize at about 10 billion. Nonetheless, feeding a population of 10 billion, with ever-increasing standards of living and change in food habits of people of emerging economies (e.g., China) will remain a major challenge until 2050 and perhaps beyond. It is in this context that the importance of sustainable management of soil resources cannot be over-emphasized.

Even though the population has ceased to double, any future increase in population (from 6.5 to 10 billion) will mostly occur in developing counties of Asia and Africa. These are also the regions where soil and water resources are already under great stress. Total demand for agricultural products by 2030 will be about 60% more than in 2000, and more than 85% of this additional demand will be in developing countries (Bruinsma, 2003). These regions are also home to most of the world's 730 million food-insecure inhabitants. Chronically food insecure people in the world were estimated at 960 million in 1970, 938 million in 1980, 830 million in 1990, 790 million in 2000, 730 million in 2005, and will be 680 million by 2010 (Rosegrant and Cline, 2003) and 440 million by 2030 (Bruinsma, 2003). The food-insecure population in 2003 included 221 million in India, 142 million in sub-Saharan Africa, 53 million in Latin America, 39 million in Near East and North Africa and 37 million in industrialized countries (FAO, 2004). The U.N. Millennium goal is to reduce world hunger by half by 2015, which will not be realized. The deficit of 13.2 million tonnes in food grains in 2000 is projected to increase to 23.3 million tonnes in 2010 (Table 1).

CLIMATE CHANGE

Global climate is affected both by natural processes and anthropogenic activities. Important among natural processes are eccentricity of Earth's orbit, solar radiation, volcanic activities, impact of meteorites, etc. The Mount Pinatubo eruption in 1991 altered the regional climate for several years. The meteorite that struck the Yucatan Peninsula 65 million years ago caused mass extinction because of change in climate. Anthropogenic activities with a strong impact on climate include those that disrupt cycles of C and H_2O, and lead to emissions of GHGs. Anthropogenic emissions of CO_2, caused by deforestation and soil cul-

TABLE 1. Estimates of food gap in developing countries (Adapted from Shapouri and Rosen, 2006).

Region	Food Deficit (Million Tonnes)	
	2000	2010
Asia	1.7	3.6
Latin America	0.6	1.0
Sub-Saharan Africa	10.7	17.5
Others	0.2	0.2
Total	13.2	22.3

tivation, probably began about 10,000 years ago (Ruddiman, 2003), and may have averted an impending ice age (Ruddiman, 2005). Since the onset of the industrial era around 1850, important anthropogenic activities responsible for increase in CO_2 emission include fossil fuel combustion, deforestation, biomass burning, soil cultivation, cement manufacturing, etc. Historic emission of CO_2 is estimated at 270 ± 30 Pg C by fossil fuel combustion and 136 ± 55 Pg by land use change and soil cultivation (IPCC, 2001). These activities have increased atmospheric concentration of CO_2 from 280 ppmv during the pre-industrial era around 1850 to 380 ppmv in 2005, and the CO_2 concentration is increasing at the rate of about 0.46% yr^{-1}. Anthropogenic activities are also increasing atmospheric concentration of CH_4 by cultivation of rice paddies, other plants (Stokstad, 2006; Keppler et al., 2006; Lowe, 2006) and livestock industry, and concentration of N_2O by use of nitrogenous fertilizers (IPCC, 2001). The observed increase in concentration of GHGs caused a global increase in temperatures by 0.6 ± 0.2°C during the 20th century, and the rate of increase was 0.17°C $decade^{-1}$ between 1950 and 2000. The observed increase in temperature caused sea level rise of 10 to 20 cm during the 20th century, increased global mean precipitation by 0.5 to 1% $decade^{-1}$, and increased the frequency of extreme events by 2 to 4%. With the doubling of CO_2 with reference to the pre-industrial level projected to be 550 ppm by 2100, the expected increase in global temperature is 2 to 4°C, precipitation of 10%, and sea-level rise of 15-75 cm (IPCC, 2001). Increase in mean temperature by 1°C in temperate regions (Northern Europe, Canada) can prolong the growing season duration by 10 days. Stabilization of atmospheric CO_2 (at 550 ppmv by 2100) would require anthropogenic emissions to be either reduced significantly, sequestered in other global pools (i.e., terrestrial, geologic, oceanic) or both. Pacala and Socolow (2004) estimated that 50 Pg of emissions would have to be sequestered by 2054. Others have estimated the sequestration to be as much as 94 Pg by 2050 and 230 Pg by 2100 (Matthews et al., 2003). The terrestrial sequestration will be an important strategy over the next two centuries, i.e., by 2200, and oceanic uptake by 2250 and beyond (Matthews et al., 2003).

SOIL CULTIVATION AND CLIMATE CHANGE

Soil cultivation is not as obvious a source of atmospheric CO_2 as is fossil fuel combustion or even deforestation. In their natural state, soils contain a large reserve of organic carbon (SOC) pool (Houghton, 2001).

The magnitude of the SOC pool depends on temperature and moisture regimes and soil profile characteristics. The SOC pool is higher in cool than in warm climates, and higher in wetter and poorly drained soils than in drier climates and well-drained soils, more in fine than in coarse-textured soils, and more in well-aggregated than poorly aggregated soils (Lal, 2001). Conversion of natural to agricultural ecosystems depletes the SOC pool by 25 to 75%. The magnitude and rate of SOC depletion are more in tropical than temperate climates, coarse-textured than heavy-textured soils, those with high than low antecedent SOC pool, and subsistence farming using extractive agricultural practices than commercial farming using science-based and recommended input (Table 2). The rate and magnitude of SOC depletion are high when input of biomass-C is lower than the loss of C from the ecosystem through erosion, leaching and mineralization or oxidation. Whereas the extractive farming practices (based on low input) lead to depletion of the SOC pool with the attendant emission of CO_2, intensive agricultural practices can also drastically impact air quality. For example, the use of nitrogenous fertilizer and the high loading of reactive nitrogen cause decreased visibility from increased aerosol production (Aneja et al., 2001). Some argue that emissions from intensive animal and crop agriculture lead to odor emissions by organic acids, reactive nitrogen in the form of ammonia and nitrogen oxides (NO_x), particulate matter from tillage and burning and H_2S (Aneja et al., 2006).

A STRATEGY FOR ACHIEVING GLOBAL FOOD SECURITY

Food security is defined as the access to food to meet the biological needs of a healthy and fulfilling life. Adoption of the green-revolution technologies increased crop yields drastically during the second half of the 20th century. The data in Table 3 show that over this 40-year period, yield increased by a factor of 2.20 for corn, 2.09 for rice, 1.55 for sorghum, 1.93 for soybean, 1.27 for sugarcane and 2.5 for wheat. Consequently, the total food production nearly tripled between 1950 and 2000, from 631 million tonnes in 1950 to 1840 million tonnes in 2000 (Table 4). However, the per capita grain consumption increased from 267 Kg in 1950 to 339 Kg in 1985 and decreased subsequently (Table 4), probably due to increase in population. Despite the unprecedented and impressive increase in food grain production, the average crop yield remains low in several developing countries. The data in Table 5 show that the highest and the lowest average yield of wheat differs

TABLE 2. Factors affecting the magnitude and rate of depletion of soil organic carbon pool upon conversion from natural to agricultural ecosystems.

Severe Depletion	Moderate or Low Depletion
1. Tropical climates	1. Temperate region
2. Extractive farming with low external input	2. Intensive agriculture with recommended level of external input
3. Biomass burning	3. Using biomass as mulch or amendment
4. Plowing and other soil disturbances	4. No-till farming without soil disturbance
5. Erosion promoting technology	5. Conservation-effective technology
6. Summer following	6. Incorporation of cover crops in the rotation cycle
7. Excessive grazing	7. Controlled grazing at low stocking rate

TABLE 3. Trends in global mean yield of principal crops (Modified from Clay, 2004).

Crop	Mean Grain Yield					Ratio of yield in 2000: 1961
	1961	1970	1980	1990	2000	
			$Mg\ ha^{-1}$			
Corn	1.94	2.35	3.16	3.68	4.27	2.20
Rice	1.87	2.38	2.75	3.54	3.90	2.09
Sorghum	0.89	1.13	1.20	1.37	1.38	1.55
Soybean	1.13	1.48	1.60	1.90	2.18	1.93
Sugarcane	50.3	54.8	55.3	61.6	64.1	1.27
Wheat	1.09	1.49	1.86	2.56	2.74	2.51

TABLE 4. Global grain prodution and per capita consumption between 1950 and 2000 (Modified from Kondratyev et al., 2003).

Year	Production (10^6 Mg)	Per Capita Consumption (Kg yr^{-1})
1950	631	267
1955	759	273
1960	824	271
1965	905	270
1970	1079	291
1975	1237	303
1980	1430	321
1985	1647	339
1990	1769	335
1995	1713	301
2000	1840	303

TABLE 5. Average yields (betwen 1996 and 2000) of wheat and corn in developed and developing countries (Recalculated from Bruinsma, 2003).

Wheat		Corn	
Country	Yield (Mg/ha)	Country	Yield (Mg/ha)
U.K.	7.8	Italy	9.4
Germany	7.3	Spain	9.1
Denmark	7.1	France	8.6
France	7.0	U.S.A.	8.2
Egypt	6.0	Canada	7.4
Hungary	3.9	Egypt	7.3
Poland	3.4	Hungary	6.0
Italy	3.2	Argentina	5.0
China	3.1	Yugoslavia	4.0
U.S.A.	2.7	China	3.8
Spain	2.6	Thailand	3.5
India	2.6	Romania	3.0
Romania	2.6	Brazil	2.6
Ukraine	2.5	Indonesia	2.6
Argentina	2.4	South Africa	2.5
Canada	2.4	India	1.7
Pakistan	2.2	Philippines	1.6
Turkey	2.1	Nigeria	1.3
Australia	2.0		
Iran	1.6		
Russia	1.4		
Kazakhstan	0.8		

by a factor of about 10 (7.8 versus 0.8 Mg ha^{-1}). Similarly, the highest and the lowest average yields of corn differ by a factor of more than 7 (9.4 versus 1.3 Mg ha^{-1}). While such a wide yield gap is indicative of the vast potential of enhancing productivity, it is important to realize that there is a wide range of interacting factors determining the national average crop yields. Thus, the data in Tables 4 and 5 must be interpreted with great caution.

Despite the difficulties of comparing national average crop yields among diverse biophysical and socio-economic conditions, such as those that affect agronomic yields in China versus India, the data in Tables 6 and 7 are extremely encouraging. For example, there is a large gap in ecologically attainable and actual yield of wheat (Table 6). In

TABLE 6. Actual and agro-ecologically attainable yield of wheat (average of 1996-2000) in some selected countries (Recalculated from Bruinsma, 2003).

Country	Ecologically Attainable Yield (Mg/ha)	Actual Yield
	- - - - - - - - - - - - - - - - - Mg/ha- - - - - - - - - - - - - - - - - -	
Germany	7.60	7.25
Poland	7.03	3.40
Denmark[1]	7.00	7.25
U.K.[1]	6.73	7.83
France[1]	6.70	7.30
Italy	6.50	3.30
Hungary	6.41	3.83
Romania	6.30	2.50
Ukraine	6.25	2.40
U.S.A.	5.84	2.83
Turkey	4.84	2.10
Russia	4.50	1.25
Canada	4.50	2.50
Argentina	4.20	2.20
Australia	4.20	2.00

[1]Actual grain yields of wheat in Denmark, U.K., and France are either close to or even higher than the attainable yields under rainfed high input farming.

TABLE 7. Actual and attainable yields of wheat and rice in India (Modified from Bruinsma, 2003).

Year	Wheat Yield	Rice Yield
	- - - - - - - - - - - - - - - - - Mg/ha- - - - - - - - - - - - - - - - - -	
1972-1974	1.26	1.63
1984-1986	1.95	2.21
1993-1995	2.48	2.83
Attainable[1]	4.00	5.40

[1]Attainable yield is weighted average of irrigated and rainfed conditions.

many countries (e.g., Italy, Hungary, Romania, Ukraine, U.S.A., Turkey, Russia, Canada, Argentina and Australia), the attainable wheat yield is almost double the actual national average yield. The data in Table 7 show an enormous potential of enhancing the yields of wheat and rice in India.

STRATEGIES FOR ENHANCING CROP YIELDS

Agricultural intensification, cultivating the best soils with best management practices to produce the optimum sustainable yield and save agriculturally marginal lands for nature conservancy, is the most feasible option of achieving global food security. The strategy is to increase production from existing croplands rather than bringing new land under cultivation (Cassman et al., 2003). Increasing irrigated land area may also be difficult (Postel, 1999). Yet, 14% more water will be needed for irrigation by 2030 than in 2000 (Bruinsma, 2003). Indeed, the goal is to set aside or retire agriculturally marginal soils. There is a strong need to critically assess technological options for sustainable management of soil and water resources, because ecosystems utilized by human societies are only sustainable in the long term if the outputs of all components produced balance the inputs into the system. Whether the required amount of plant nutrients to obtain the desired yield is supplied in organic rather than inorganic form is a matter of availability and logistics. Plants cannot differentiate the nutrients supplied through the organic or inorganic sources. An objective question is of nutrient availability, in sufficient quantity, in appropriate form, and at the time the nutrients are needed for the optimum growth and yield formation. The goal is to achieve a positive and balanced application of nutrients through integrated nutrient management (INM) (Sanchez, 2002). The latter involves a judicious combination of chemical fertilizers supplemented with organic manures and biologic nitrogen fixation (BNF). The use of organic amendments can enhance soil quality through increase in the SOC pool and the microbial biomass C. Despite the benefits of using manures and organic amendments, there are no viable alternatives to using chemical fertilizers to meet the food needs of 10 billion inhabitants on 1600 Mha of the existing area of arable cropland. The global fertilizer demand will continue to increase (IFDC, 2004). Therefore, recycling nutrients by returning crop residues to soil as mulch or an amendment (compost) is important. Removing crop residues for fodder or biofuels (Glassner et al., 1998; Sokhansanj et al., 2002; Wolf et al., 2003; Lal, 2005) can jeopardize food security and environmental quality.

The data in Table 8 show the global average cereal grain yield of 2.64 tonnes ha^{-1} and cereal production of 1267 million tonnes during 2000. To maintain the same caloric input and the diet habit, the grain yield and production will have to be increased, respectively, to 3.60 Mg ha^{-1} and 1706 million tonnes (35% increase) by 2025 and 4.30 Mg ha^{-1} and 1995 million tonnes (58% increase) by 2050. With a likely change in

TABLE 8. Average yield of cereals required in developing countries[1] by 2025 and 2050 to meet the required production with no increase in area (Adapted from Wild, 2003).

Parameter	Cereal Grain Yield (Mg/ha)	Cereal Production(10^6 Mg/yr)
1. Present (2000)	2.64	1267
2. Required		
a. 2025		
(i) + 35%[2]	3.60	1706
(ii) + 62%	4.40	2045
b. 2050		
(i) + 58%[3]	4.30	1995
(ii) + 121%	6.00	2786

[1]Africa, South America, Asia (excluding Japan).
[2]35% and 62% increase above the present level in 2025 estimated to account for increase in population and change in food habits.
[3]58% and 121% increase above the present level in 2050 and estimated to account for increase in population and change in food habits.

diet of the population in emerging economies (e.g., China and India), however, grain yield and production will have to be increased, respectively, to 4.40 tonnes ha^{-1} and 2045 million tonnes (62% increase) by 2025 and 6.00 tonnes ha^{-1} and 2786 million tonnes (121% increase) by 2050.

TECHNOLOGICAL OPTIONS
FOR AGRICULTURAL INTENSIFICATION

The overall strategy is to intensify agriculture by adopting 'Recommended Management Practices' (RMPs) and land-saving technologies. The strategy is to: (i) adopt no-till farming, (ii) recycle crop residue as surface mulch, (iii) create positive nutrient balance through judicious combination of manuring and fertilizers along with biological nitrogen fixation, (iv) deliver nutrients and water through sub-fertigation and precision farming, and (v) adopt complex rotations involving cover crops and without summer fallow (Table 9). The schematics in Figure 3 outline the basic concept of delivering water and nutrients directly to plant roots. Widespread adoption of RMPs will enhance soil quality, conserve soil and water, improve biodiversity, and reduce risks of water

TABLE 9. Recommended agricultural practices for agricultural intensification.

Management System	Recommended Agricultural Practice
1. Seed bed preparation	Conservation tillage or no-till farming.
2. Moisture conservation	Mulch farming, elimination of summer fallow.
3. Rotation	Complex rotations with cover crops.
4. Soil fertility management	Integrated nutrient management for positive nutrient balance of N, P, K, Ca, Mg, Cu, Zn, Mo, etc.
5. Irrigation	Sub-irrigation, drip irrigation, fertigation.
6. Subsistence farming	Commercial, intensive agriculture.
7. Marginal soil	Set aside for nature conservancy.
8. Natural resources management	Integrated watershed management.
9. Residue management	Retention as surface mulch (no removal or burning).
10. Drainage of wetlands	Restore drained peat soils.

FIGURE 3. Delivering nutrients (integrated nutrient management systems) and water (drip irrigation) directly to roots of genetically modified or improved plants through soil-specific management (or precision farming) in conjunction with no-till method of seedbed preparation, use of crop residue mulch, and integrated pest management techniques (Modified from Lal, 2000).

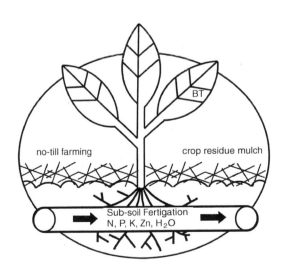

pollution and contamination. A relevant example is the severity and extent of soil erosion in the USA. It was estimated that 91 Mha of agricultural land had been devastated by severe erosion during the 1930s (Utz et al., 1938). Yet, the land area affected by severe erosion during the 1990s was only 24 Mha, because most eroded land had been restored. Soil conservation in the USA has been a global success story (Trimble, 1999), as was the case in the highlands of Machakos, Kenya (Tiffen et al., 1994). Therefore, a solution to the severe problem of soil degradation in developing countries (Table 10) is conversion to an appropriate land use and adoption of RMPs for agricultural intensification on arable lands. The area under no-till farming is less than 5% (90 Mha out of 1600 Mha) of the global cropland area (Table 11). Increase in area under no-till farming, as has been the case in Latin America (e.g., Brazil, Argentina, Chile), would sustain production and improve quality of soil and water resources.

SOIL CARBON SEQUESTRATION FOR ADVANCING FOOD SECURITY AND MITIGATING CLIMATE CHANGE

There are a wide range of RMPs that can improve soil quality and advance global food security, and provide an evolutionary step in improving agricultural technologies over the next 25 years and projected for the 21st century (Figure 4). The data in Table 8 show that a quantum jump in food production is needed. Appropriate technologies to bring about this quantum jump include (i) improved soil, water and nutrient management to enhance use efficiency of off-farm inputs, (ii) restoration of soils degraded by erosion, salinization, nutrient depletion and compaction, (iii) using integrated nutrient management and precision farming to reduce losses and enhance use efficiency, (iv) adopt no-till

TABLE 10. Extent of soil degradation in developing countries (Modified from Oldeman, 1994).

Degradative Process	Area Affected 10^6 Ha	
	Developing Countries	World
Water erosion	837	1094
Wind erosion	455	548
Chemical degradation	213	240
Physical degradation	44	83

TABLE 11. Cropland under no-till farming in key countries in 2003-06 (Adapted from Brown, 2004).

Country	Area Under No-Till (10^6 Ha)
United States	23.7
Brazil	21.9
Argentina	16.0
Canada	13.4
Australia	9.0
Paraguay	1.5
Pakistan/India	1.5
Bolivia	0.4
South Africa	0.3
Spain	0.3
Venezuela	0.3
Uruguay	0.3
France	0.2
Chile	0.2
Others	1.2
Total	90.1

farming with crop residue mulch and cover crops, and (v) use diverse farming systems, including agroforestry.

Adoption of these technologies would restore the depleted SOC pool and enhance soil quality (Figure 5). The rate of SOC sequestration may range from 50 Kg ha^{-1}yr^{-1} in arid tropics to 1000 Kg ha^{-1}yr^{-1} in humid temperate regions (Lal, 2001; 2004b). The SOC sequestration has potential to sequester 1 Pg C yr^{-1} by 2054 through a widespread adoption of no-till farming (Pacala and Socolow, 2004). Land C sinks are important to mitigating the climate change (Royal Society of London, 2000). The SOC sequestration, an ancillary benefit of agricultural intensification, can reduce the rate of increase of atmospheric CO_s concentration and increase agronomic productivity.

Results of several experiments, conducted in diverse agro-ecoregions, show increase in agronomic yields by improvement in the SOC pool. The data in Table 12 show that increasing the SOC pool in the root zone by 1 Mg C ha^{-1}yr^{-1} through adoption of RMPs can increase grain yield of corn by 100 to 300 Kg ha^{-1}, soybeans by 20-50 Kg ha^{-1}, wheat by 20-70 Kg ha^{-1}, rice by 10-50 Kg ha^{-1}, and beans by 30-60 Kg ha^{-1}. Therefore, a widespread adoption of RMPs on a global scale can increase food grain production by 32 ± 11 million tonnes yr^{-1} (Table 13), and eliminate food deficit in sub-Saharan Africa, South Asia and other

FIGURE 4. Technological innovations for agricultural intensification to meet food demands of population projected to increase from 6 billion in 2000 to 10 billion by 2100 (Modified from Lal, 2000).

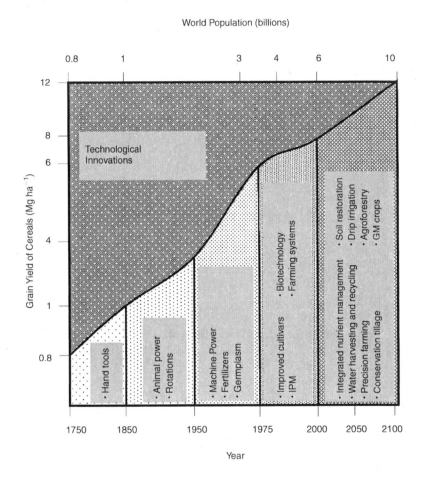

global hot spots where food insecurity has plagued the human society for decades if not centuries.

It is in this regard that agricultural intensification through adoption of RMPs is a truly win-win situation. It will achieve food security, save land for nature conservancy and improving biodiversity, and offset CO_2 emissions by SOC sequestration. Nonetheless, impacts of agricultural intensification on air quality and non-point source pollution must not be overlooked (Krupa, 2003; NRC, 2003; Aneja, 2006). Anoxia in the Gulf

FIGURE 5. Depletion of soil organic carbon pool upon conversion from natural to agricultural ecosystems is caused by mineralization, leaching and erosion. The sink capacity thus created can be filled by soil C sequestration through adoption of recommended management practices (e.g., no-till, mulching, manuring, using cover crops and agroforestry, etc.). The attainable soil C sink capacity is generally two-thirds to three-fourths of the potential sink capacity.

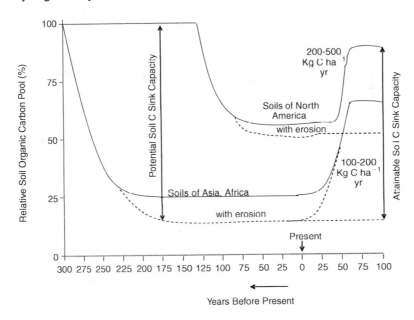

of Mexico is caused by the nitrogen losses from cropland in the U.S. Midwest (Goolsby et al., 2001). No-till farming causes more volatilization losses of NH_3 and N_2O emission than conventional tillage (Venterea et al., 2005). The scientific knowledge about nitrogen, volatile organic compounds, sulfur and particulate matter emissions from intensively managed agriculture and the fate of these compounds is scanty. The impact of intensive agricultural practices (crops and livestock) on quality of air and water must be assessed to develop appropriate guidelines for minimizing the adverse impact.

CONCLUSION

For both issues, global food security and climate change, a solution lies in adopting RMPs for sustainable management of soil and water re-

TABLE 12. Crop yield increase by increase in soil organic carbon by 1 Mg ha^{-1}yr^{-1} (Adapted from Lal, 2006).

Crop	Yield Increase (Kg ha^{-1}yr^{-1})
Maize	100-300
Soybeans	20-50
Wheat	20-70
Rice	10-50
Sorghum	80-140
Millet	30-70
Beans	30-60

TABLE 13. Projected increase in food grain production in developing countries by increase in soil organic carbon pool by 1 Mg ha^{-1}yr^{-1} (Adapted from Lal, 2006).

Region	Production Increase (10^6 Mg yr^{-1})
Africa	4.6 ± 1.6
Latin America	8.4 ± 2.4
Asia	18.64 ± 7.1
Total	31.6 ± 11.1

sources. While achieving food security, adoption of RMPs would also restore degraded soils and ecosystems, and improve water quality. With the increase in world population from 6.5 billion in 2006 to 10 billion toward the second half of the 21st century, the per capita land area will decrease and natural resources will be under more stress than ever before. Therefore, it is important to adopt the land saving technologies of agricultural intensification. World soils have the capacity to feed the present and future population provided that soils are used, improved and restored. Soils have also the capacity to sequester atmospheric C at the rate of about 1 Pg yr^{-1} by 2050 through adoption of no-till farming and other RMPs. Contrary to the misconception, adoption of RMPs is a solution to the environmental issues and also to achieving global food security. The history of vanished civilizations has taught us that for the modern civilization to thrive and flourish, its motto must be "In Soil We Trust."

REFERENCES

Aneja, V. P., P. A. Roelle, G. C. Murray, J. Southerland. J. W. Erisman, D. Fowler, W. A. H. Asman and N. Patni, 2001. Atmospheric nitrogen compounds. II. Emissions, transport, transformation, deposition and assessment. Atmos. Environ. 35: 1903-1911.

Aneja, V. P., W. H. Schlesinger, D. Niyogi, G. Jennings, W. Gilliam, R. E. Knighton, C. S. Duke, J. Blunden and S. Krishnan, 2006. Emerging national research needs for agricultural air quality. EOS 87(3): 25, 29.

Brown, L. R., 2004. Outgrowing the Earth: The Food Security Challenge in an Age of Falling Water Tables and Rising Temperatures. W. W. Norton and Co., 239 pp.

Bruinsma, J., (Ed.) 2003. World Agriculture: Towards 2015/2020. FAO/Earthscan, London, 432 pp.

Cassman, K. G., A. Doebermann, D. T. Walters and H. Yang, 2003. Meeting cereal demand while protecting natural resources and improving environment quality. Ann. Rev. Environ. Resour. 28: 315-358.

Clay, J., 2004. World Agriculture and the Environment. A Commodity by Commodity Guide to Impacts and Practices. Island Press, Washington, DC, 570 pp.

Cohen, J. E., 2003. The human population: Next half century. Science 302: 1172-1175.

Dailey, G. C., 1995. Restoring value to the world's degraded lands. Science 269: 350-354.

Diamond, J., 2005. Collapse. How Societies Choose to Fail or Succeed. Viking, New York.

FAO, 2004. FAO and the Challenge of the Millennium Development Goals: The Road Ahead. FAO, Rome, Italy, 39 pp.

Glassner, D. A., J. R. Hettenhaus and T. M. Schechinger, 1998. Corn stover potential: A scenario that can recast the corn sweetener industry. In J. Janick (Ed.), Perspective on New Crops and New Uses. Proc. IVth Natl. Symp.: New Crops and New Uses. Biodiversity and Agric. Sustainability, Phoenix, AZ. ASHA Press, Alexandria, VA.

Goolsby, D. A., W. A. Battaglin, B. T. Aulenbach and R. P. Hooper, 2001. Nitrogen input to Gulf of Mexico. J. Environ. Qual. 30: 329-336.

Houghton, R. A., 2001. Counting terrestrial sources and sinks of carbon. Climatic Change 348: 525-534.

IFDC, 2004. Global and regional data on fertilizer production and consumption, 1961/62-2002/03. IFDC, Muscle Schoals, AL, 73 pp.

IPCC, 2001. Climate Change 2001: The Scientific Basis. Cambridge Univ. Press, Cambridge, UK, 880 pp.

Keppler, F., J. T. G. Hamilton, M. Brab and T. Rockmann, 2006. Methane emissions from terrestrial plants under aerobic conditions. Nature 439: 187-191.

Kondratyev, K. Y., V. F. Krapivin and C. A. Varotsos, 2003. Global Carbon Cycle and Climate Change. Springer-Verlag, Berlin, 368 pp.

Krupa, S. V., 2003. Effects of atmospheric ammonia (NH_3) on terrestrial vegetation: Review. Environ. Pollut. 124: 179-221.

Lal, R., 2000. Controlling greenhouse gases and feeding the globe through soil management. Univ. Distinguished lecture, 17 Feb. 2000, Wexner Center, The Ohio State University, Columbus, OH 43210.

Lal, R., 2001. World cropland soils as a source of sink for atmospheric carbon. Adv. Agron. 71: 145-191.

Lal, R., 2004a. Carbon emission from farm operations. Env. Intl. 30: 981-990.

Lal, R., 2004b. Soil carbon sequestration impacts on global climate change and food security. Science 304: 1623-1627.

Lal, R., 2005. World crop residue production and implications of its use as biofuel. Env. Intl. 31: 575-584.

Lal, R., 2006. Enhancing crop yields in the developing countries through restoration of soil organic carbon pool in agricultural lands. Land Degrad. & Rehab. (In Press.)

Lowe, D. C., 2006. A green source of surprise. Nature 439: 148-149.

Matthews, H. D., A. J. Weaver, M. Eby and K. J. Meissner, 2003. Radiative forcing of climate by historical land cover change. Geophysical Research Letters 30(2), 1055 doi:10 1029/202 GL016098.

NRC, 2003. Air emissions from animal feeding operations, current knowledge, future needs. National Research Council, National Academy Press, Washington, DC, 263 pp.

Oldeman, L. R., 1994. The global extent of soil degradation. In D. J. Greenland and I. Szabolcs (Eds), Soil Resilience and Sustainable Land Use, CAB International, Wallingford, UK, 99-118.

Pacala, S. and R. Socolow, 2004. Stabilization wedges: Solving the climate problem for the next 50 years with current technologies. Science 305: 968-972.

Postel, S., 1999. Pillar of Sand, Can the Irrigation March Last? W. Norton and Co., NY, 313 pp.

Rees, W. F., 2004. The eco-footprint of agriculture: A far from (thermodynamic) equilibrium interpretation. In NABC Report 16 Agricultural Biotechnology: Finding Common International Goals. National Agric. Biotech. Council, Ithaca, NY: 87-110.

Rosegrant, M. W. and S. A. Cline, 2003. Global food security: Challenges and policies. Science 302: 1917-1919.

Royal Society of London, 2000. The role of land carbon sinks in mitigating global climate change. The Royal Society, London, 27 pp.

Ruddiman, W. F., 2003. The anthropogenic greenhouse era began thousands of years ago. Climate Change 61: 261-293.

Ruddiman, W. F., 2005. How did humans first alter global climate. Scientific American 292: 429-436.

Sanchez, P. A., 2002. Soil fertility and hunger in Africa. Science 295: 2019-2020.

Scherr, S. J. and S. Yadav, 1999. Soil degradation. A threat to developing country's food security. IFPRIU Discussion Paper 27, Washington, DC, 63 pp.

Shapouri, S. and S. Rosen, 2006. Soil degradation and food aid needs in low-income countries. In R. Lal (Ed.), Encyclopedia of Soil; Science. 2nd Edition, Taylor and Francis, Boca Raton, FL, 425-427.

Sokhansanj, S., A. Turhollow, J. Cushman, J. Cundiff, 2002. Engineering aspects of collecting corn stover for bioenergy. Biomass and Bioenergy 23: 347-355.

Stokstad, E., 2006. Plants may be hidden methane source. Science 311: 159.

Tiffen, M., M. Mortimore and F. Gichucki, 1994. More People, Less Erosion: Environmental Recovery in Kenya. J. Wiley & Sons, Chichester, UK, 310 pp.

Trimble, S. W., 1999. Decreased rates of alluvial sediment storage in the Coon Creek Basin, Wisconsin 1975-93. Science 285: 1244-1247.

Utz, E. J., C. E. Kellog, E. H . Reed, J. H. Stallings and E. N. Munns, 1938. The problem: The nation as a whole. In Soils and Men, Yearbook of Agriculture, 1938, Washington, DC, 84-110.

Venterea, R. T., M. Burger and K. A. Spokas, 2005. Nitrogen oxide and methane emissions under varying tillage and fertilizer management. J. Environ. Qual. 34: 1467-1477.

Vitousek, P. M., H. A. Mooney, J. Lubachenco, and J. M. Melillo, 1997. Human domination of Earth's ecosystems. Science 277:494-499.

Wild, A., 2003. Soils, Land and Food: Managing the Land during 21st Century. Cambridge Univ. Press, Cambridge, UK, 245 pp.

Wolf, J., P. S. Bindraban, J. C. Luijten and L. M. Vleeshouwers, 2003. Exploratory study on the land area required for global food supply and the potential global production of bioenergy. Agric. Systems. 76: 841-861.

doi:10.1300/J411v19n01_03

Whole-System Integration and Modeling Essential to Agricultural Science and Technology for the 21st Century

L. R. Ahuja
A. A. Andales
L. Ma
S. A. Saseendran

SUMMARY. In the 21st century, agricultural research has more difficult and complex problems to solve. The continued increase in population in the developing countries requires continued increases in agricultural production. However, the increased use of fertilizers, pesticides, and water required for the new higher yielding crop varieties has been causing environmental problems. Excessive leaching and runoff of agricultural chemicals are seriously affecting the quality of both the groundwater and surface waters. Increase in soil salinity, decline in soil organic matter, and increase in soil erosion remain the major problems in intensively farmed areas. Even the air quality is being affected. At the same time, market-based global competition is challenging the eco-

L. R. Ahuja, A. A. Andales, and L. Ma are affiliated with the USDA-Agricultural Research Service-Agricultural Systems Research Unit, 2150 Centre Avenue, Building D, Suite 200, Fort Collins, CO 80526 USA (E-mail: laj.ahuja@ars.usda.gov).

S. A. Saseendran is affiliated with Colorado State University, Fort Collins, CO 80523 USA.

[Haworth co-indexing entry note]: "Whole-System Integration and Modeling Essential to Agricultural Science and Technology for the 21st Century." Ahuja, L. R. et al. Co-published simultaneously in *Journal of Crop Improvement* (Haworth Food & Agricultural Products Press, an imprint of The Haworth Press, Inc.) Vol. 19, No. 1/2 (#37/38), 2007, pp. 73-103; and: *Agricultural and Environmental Sustainability: Considerations for the Future* (ed: Manjit S. Kang) Haworth Food & Agricultural Products Press, an imprint of The Haworth Press, Inc., 2007, pp. 73-103. Single or multiple copies of this article are available for a fee from The Haworth Document Delivery Service [1-800-HAWORTH, 9:00 a.m. - 5:00 p.m. (EST). E-mail address: docdelivery@haworthpress.com].

Available online at http://jcrip.haworthpress.com
doi:10.1300/J411v19n01_04

nomic viability of traditional agricultural systems. Global climate change will pose additional challenges. The solution or mitigation of these changing and multiple problems will require continual improvement or changes in management and selection of dynamic cropping systems using a whole-system approach. Therefore, synthesis and quantification of disciplinary knowledge at the whole-system level is essential to meeting these challenges. The process-based models of agricultural systems provide such a synthesis and quantification for evaluating the effects of varying management practices, crops, soils, water, and climate on both the production and the environment. These system models will greatly enhance the efficiency of field research for developing sustainable agricultural systems, serve as guides for planning and management, and help transfer new technologies to various conditions of developing countries. Current state of the system models and their applications for these purposes are reviewed, and advancements needed in models to improve and extend these applications are presented. doi:10.1300/J411v19n01_04 *[Article copies available for a fee from The Haworth Document Delivery Service: 1-800-HAWORTH. E-mail address: <docdelivery@haworthpress.com> Website: <http://www.HaworthPress.com> © 2007 by The Haworth Press, Inc. All rights reserved.]*

KEYWORDS. Synthesis of knowledge, synthesis across space and time, system approach, system models in research, decision support system, technology transfer

INTRODUCTION

Understanding real-world situations and solving significant agronomic, engineering, and environmental problems require process-based synthesis and quantification of knowledge at the whole-system level. In the 20th Century, we made tremendous advances in discovering fundamental principles in different scientific disciplines using reduction methods, which created major breakthroughs in management and technology for agricultural systems. However, as we enter the 21st Century, agricultural research has more difficult and complex problems to solve. The environmental consciousness of the general public is challenging producers to modify farm management to protect water, air, and soil quality, while staying economically profitable. At the same time, market-based global competition in agricultural production and the global climate change are threatening economic viability of the traditional agricultural systems, and require the development of new and dynamic production systems. Site-specific, optimal management of spatially variable soil, appropriately selected crops, and available water resources on the landscape can help achieve both environmental and production objectives. Fortunately, the new electronic technologies can provide a vast amount of real-time information about soil and crop conditions via remote sensing with satellites or ground-based instruments, which, combined with near-term weather, can be used to develop a whole new level of site-specific management. However, we need the means to assimilate this vast amount of data. A synthesis and quantification of disciplinary knowledge at the whole-system level, via process-based modeling of agricultural systems, are essential to develop such means and the management systems that can be adapted to continual change. Interactions among disciplinary components of the agricultural systems are generally very important. Models are the only way to find and understand these interactions in a system, integrate various experimental results and observations for different conditions, and extrapolate limited experimental results to other soil and climate conditions.

System modeling has been a vital step in many scientific disciplines. We would not have gone to the moon successfully without the combined use of good data and models. In designing of automobiles and airplanes, computer models of the system are increasingly replacing the scaled physical models of the past. Models have also been used extensively in designing and managing water resource reservoirs and distribution systems, and in analyzing waste disposal sites. Although a lot more work is needed to bring agricultural system models to the level of physics and hydraulic system models, the agricultural system models have matured enough (Ahuja et al., 2002; Matthews and Stephens, 2002; Struif Bontkes and Wopereis, 2003) that, with some good data to serve as reference, they can be used for many practical applications in research and management. These applications will expose knowledge or conceptual gaps in the models that will then be filled in time.

HOW WILL MODELS BENEFIT FIELD RESEARCH?

An agricultural system involves complex interactions among several different components and factors. Figure 1 illustrates some of these interactions of multiple factors operating through their connection to soil water. These interactions need interdisciplinary field research and quantification with the help of conceptual and process models.

FIGURE 1. An illustration of complexity of the agricultural systems via interactions among multiple components and factors through their connection to soil water.

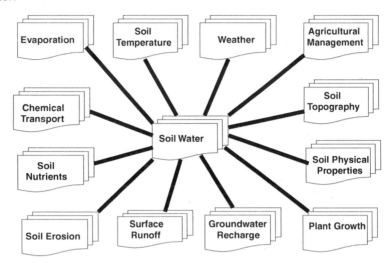

Integration of system models with field research has the potential to significantly enhance efficiency of agricultural research and raise agricultural science and technology to the next higher level. The integration will benefit both field research and models in the following ways:

- Promote a systems approach to field research that examines all component interactions,
- Facilitate better understanding of the cause-and-effect relationships in and quantification of experimental results,
- Promote efficient and effective transfer of field research results to different soil and weather conditions, and to different cropping and management systems outside the experimental plots,
- Help field researchers focus on the identified fundamental knowledge gaps and make field research more efficient, and
- Provide the needed field test and improvement of the models before delivery to other potential users–agricultural consultants, farmers, ranchers, extension agencies, and action agencies (NRCS, EPA, and others).

Modeling of agricultural management effects on soil-plant-atmosphere properties and processes has to be a centerpiece of an agricultural system model if it is to have useful applications in field research and management. An example is the ARS Root Zone Water Quality Model (RZWQM), a process level model built to simulate management effects on water quality and crop production (Figure 2; Ahuja et al., 2000). Most widely used crop models, such as the CERES and CROPGRO family, APSIM, GOSSYM, GLYCIM, EPIC/ALMANAC, need to be enhanced for simulating management effects on both production and environmental quality.

APPLICATIONS IN TECHNOLOGY TRANSFER AND MAKING MANAGEMENT AND POLICY DECISIONS

A field-tested model can be used to transfer the results of experimental research to other soil types, climates, and management conditions outside the experimental stations. It can also be used for extrapolating experimental results from a limited period of experimentation to variability in climatic conditions across longer periods (e.g., 25-100 years), and to extreme climatic conditions (e.g., droughts or flooding) not encountered during the study period. A validated model is an excellent tool for in-depth analysis of problems in management, environmental

FIGURE 2. Processes and time steps in RZWQM. Management practices are the centerpiece of this process-based cropping system model. (Adapted from Ahuja et al., 2000.)

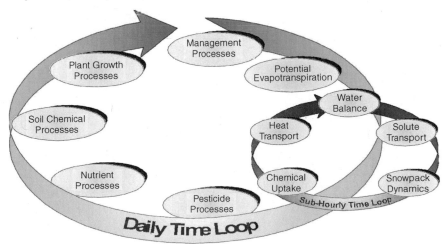

quality, global climate change, and other new emerging issues. It can thus be a basis for national policies. Models can also be used to explore new ideas and strategies under different weather and climatic conditions before testing them in the field.

A field-tested model can also be used as a decision aid in choosing the best management practices for sustainable production over the long term (e.g., Andales et al., 2003), as well as to guide site-specific management on variable landscapes and within-season dynamic management in response to variable soil moisture and weather conditions (e.g., Ahuja and Ma, 2002). The new decision support systems (DSSs) have an agricultural system model at their core, but are supported by soil, climate, and management databases, environmental and economic analysis packages, user-friendly interfaces to check the default data and enter site-specific data, and a graphical visualization of results. An example is the design of USDA-ARS, GPFARM-DSS (Figure 3; Ascough et al., 1995; Andales et al., 2003). The GPFARM (Great Plains Framework for Agricultural Resource Management) is a whole-farm decision support system for strategic planning and evaluation of alternate cropping systems, range-livestock systems, and integrated crop-livestock farming options, for production, economics, and environmental impacts.

Currently, process level models may be difficult for agricultural consultants, extension field office personnel, and producers to use. A new approach toward a DSS is to create an integrated research-information database as a core of the DSS in place of a model. A system model, validated against available experimental data, is used to generate production and environmental impacts of different management practices for all major soil types, weather conditions, and cropping systems outside the experimental limits. This model-generated information is then combined with experimental data and the long-term experience of the farmers and field professionals to create a database (Rojas et al., 2000). The database can be combined with an economic analysis package. It may also be connected to a so-called "Multi-Objective DSS" for conducting a tradeoff analysis between conflicting objectives, such as economic return and environmental quality (Heilman et al., 2002). It is also very flexible in generating site-specific recommendations.

The most desirable vision for agricultural research and technology is to have a continual, two-way interaction among cutting-edge field research, conceptual and process-based models, and DSSs (Figure 4; Ahuja et al., 2002).

FIGURE 3. The design of GPFARM decision support system (DSS). (Adapted from Ascough et al., 1995 and Andales et al., 2003.)

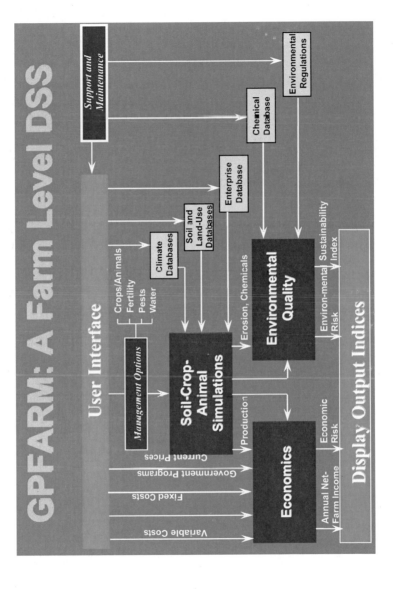

FIGURE 4. Interactions among filed research, process-based system models, and decision support systems. (Adapted from Ahuja et al. 2002.)

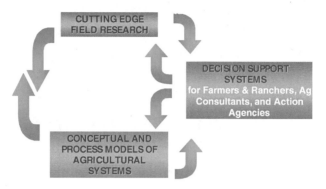

EXAMPLES OF SYSTEM MODEL APPLICATIONS

In this section, we provide a broad sampling of applications of system models in research and strategic management. More comprehensive reviews of system-model applications are provided by Ahuja et al. (2002), Matthews and Stephens (2002), and Struif Bontkes and Wopereis (2003), among others. Several examples were taken from these three compilations while others were selected from the literature.

Enhanced Understanding of the Complexities in Experimental Data

The field experimental results are most often the outcome of the complex effects and interactions of several soil, climatic, and biological factors. As a consequence, the results vary from location to location and year to year. The cause-and-effect relationships in the data are extremely difficult to discern. The models are built upon the concepts and theories developed from the past hypothesis-based experimental research and include the various interactions to the extent of current knowledge. The models should account for the varying soils and climatic conditions for different locations and years for model simulations of the data to help explain variation of results and cause-and-effect relations in the results. When a model is tested against a good quality dataset and the model results deviate significantly from the experimental results, it may point to a possible knowledge gap in the model. Matthews and Stephens (1998) provided a good example of this. An initial model for tea (*Camellia sinensis*) showed that temperature alone could not be used to simulate a large peak in tea production in September of each year in Tanzania. A mechanism that could explain the peak was the assumption that the growth of dormant shoots was triggered at the time of winter solstice, allowing a large number of shoots to develop simultaneously and reach harvestable size at the same time. The proposed mechanism was based on the hypothesis that shoot dormancy was induced by declining photoperiod and released by increasing photoperiod. Additional research was needed to test this hypothesis more carefully, but the model did help enhance understanding of the mechanism.

Another example is provided in the area of chemical transport in soils. In certain soils, the distribution of a non-adsorbed chemical deviated from the Gaussian distribution that is theoretically expected and commonly found (Figure 5, top; Ahuja et al., 1995), in that the chemical seemed to be retained longer near the soil surface. Experiments and modeling in soil columns showed that when the surface soil layer consisted of stable aggregates, more chemical was retained near the soil surface (Figure 5, bottom; Ahuja et al., 1995).

Management of Crop Production

It is both difficult and impractical to conduct plot–or field–scale experiments that consider all possible management options across numerous environments. An obvious application of cropping system models is to simulate various management options in different environments to predict effects on crop production. Through careful interpretation of simulation results, researchers can identify the best management practices under the simulated conditions. In this section, applications of cropping-system models in studying the effects of various management practices on yield are highlighted.

Crop models have been used in yield-gap analyses to quantify potential yield compared with actual yield. Models can be used to estimate potential yields at multiple locations to determine effects of genotype and the environment (e.g., soils, climate, and day length). In one such example, Aggarwal et al. (1995) used the WTGROWS model to predict potential wheat yields across a latitude gradient in India. They determined that the yield gap was at least 2000 kg ha^{-1}, which indicated that much more could be done in wheat management to increase actual yields. A major finding in the study was that late sowing was contributing significantly to the yield gap.

FIGURE 5. Observed and model-simulated results of bromide movement in soil: top–the commonly expected Gaussian distribution of the chemical pulse; bottom–soil aggregates at the surface keep more chemical near the surface. (Adapted from Ahuja et al., 1995.)

Bromide Concentration in Soil Water (μg/ml)

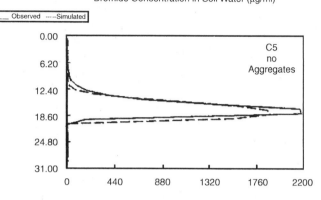

Initially dry conditions with no macropores

Bromide Concentration in Soil Water (μg/ml)

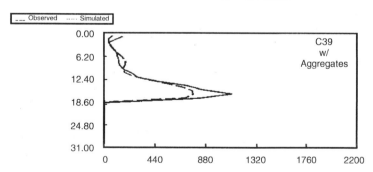

Initially dry conditions with no macropores

Some crop-specific models can be used to simulate differences in yield among varieties. This is achieved by using the appropriate genetic coefficients in the model that characterize each variety. Dimes et al. (2003) used such an approach with the APSIM model in Zimbabwe. Using simulations of maize growth and yield, they demonstrated to local farmers the impact of changing from a short-duration variety commonly used in the area to a medium-duration variety. An 11-year simulation showed that this would have a detrimental effect on maize yields in most years, and local farmers clearly appreciated this information.

Oftentimes, there is interest in studying the interactions between variety and one or more management practices, such as planting date or nitrogen (N) fertilization. One example is the study by Dzotsi et al. (2003) using the Decision Support System for Agrotechnology Transfer (DSSAT) model to derive optimum combinations of cultivar and sowing dates for maize in Southern Togo. The study quantified the risks (standard deviations of yield) associated with combinations of varieties and planting dates. From the simulated data, informational leaflets were made and distributed to farmers as a guide in the selection of maize variety as a function of the preferred time of sowing.

Some crop models do not simulate all the major plant processes but rather focus on a single aspect or process, such as phenology. These models, as well as the more detailed models that simulate assimilation and allocation to plant parts, can be used to develop cropping calendars that give recommended dates for different operations. For example, the Rice Development (RIDEV) phenology model was used to simulate rice phenology for 30 years of historical weather data in Podor, Senegal to develop a cropping calendar (Wopereis et al., 2003). The cropping calendar included estimated dates for sowing, transplanting, split-applications of urea, last drainage, and harvest (Table 1).

A common application of crop models is the simulation of interactions between crop yield and levels of agricultural inputs such as irrigation water or nitrogen (N) fertilizer. Because models keep track of the water balance and N amounts in the soil profile as well as the estimated uptake by crops, the models can help estimate the proper amounts and timing of irrigation or N fertilization.

Saseendran et al. (2004) used RZWQM to quantify interactions between N level and wheat yield at Akron, Colorado (Figure 6). They found that the model was sensitive enough to simulate differences in biomass and grain yield among different N application rates. Thus, the model could potentially be used to determine N application rates, given soils and weather information.

Using data from the same location, Ma et al. (2000) used RZWQM to simulate soybean grain yield under different levels of irrigation (Figure 7). The study showed that the model had adequate sensitivity to correctly simulate the response of soybeans to various levels of water availability.

TABLE 1. RIDEV estimated cropping calendars using 7-day intervals for transplanted rice, cultivar Jaya, during the wet season in Podor, Senegal. (Based on simulations using 30 years of historical weather data.) (Modified from Wopereis et al., 2003.)

Trans-planting Date	First Urea Split	Second Urea Split	Third Urea Split	Date of Flowering	Date of Last Drainage	Harvest Date	Cropping Cycle (days)
July 13	July 31	Aug 27	Sept 16	Sept 26	Oct 11	Oct 25	125
July 20	Aug 7	Sept 2	Sept 22	Oct 2	Oct 17	Oct 31	124
July 27	Aug 14	Sept 8	Sept 29	Oct 9	Oct 23	Nov 6	125
Aug 3	Aug 21	Sept 14	Oct 5	Oct 15	Oct 29	Nov 12	125
Aug 10	Aug 28	Sept 21	Oct 11	Oct 21	Nov 5	Nov 19	125
Aug 17	Sept 4	Sept 28	Oct 18	Oct 28	Nov 12	Nov 26	126
Aug 24	Sept 11	Oct 5	Oct 26	Nov 5	Nov 19	Dec 3	128
Aug 31	Sept 18	Oct 13	Nov 3	Nov 13	Nov 27	Dec 11	131
Sept 7	Sept 25	Oct 23	Nov 12	Nov 22	Dec 7	Dec 21	135
Sept 14	Oct 2	Nov 2	Nov 22	Dec 2	Dec 17	Dec 31	140
Sept 21	Oct 9	Nov 14	Dec 5	Dec 15	Dec 29	Jan 12	147
Sept 28	Oct 16	Nov 28	Dec 19	Dec 29	Jan 12	Jan 26	154

The CERES-Maize model was used in Malawi to perform modeling experiments to investigate the N fertilizer response on different soils (Singh et al., 2002). The complex interaction between soil moisture and nutrient rates across 25 seasons is shown in Figure 8.

Ahuja and Ma (2002) provided an example of using RZWQM to link irrigation timing and amount to soil water contents in the root zone. The model was used to simulate maize silage yields, N uptake by the crop, and N leaching at different irrigation amounts (Figure 9) at a site in eastern Colorado. The lower limit of available water to trigger irrigation was varied from 10% to 90%, and the upper limit to stop irrigation from 50% to 90%. The simulations showed that irrigations should be triggered when available water is about 20% and stopped when soil water storage is filled to about 50% to maximize N uptake and silage yield and minimize nitrate-N leached.

The model application described by Ma et al. (2000) is a good example of using RZWQM to determine the best time of nitrate-N application to optimize N uptake and silage yield while minimizing nitrate-N leaching (Figure 10). Their finding was that early-season was the best time to apply N.

Carberry et al. (2004) used simulation modeling in a novel way by combining a participatory research approach (direct interaction with

FIGURE 6. Field-measured and RZWQM model-predicted winter wheat grain and biomass yields during three crop seasons (1987-98, 1988-89 and 1989-90) under different
N nutrient rates (0 kg N ha^{-1} of 1987-88 was used for calibration of the model). Error bars
represent one standard deviation of the measured grain yield from the mean.

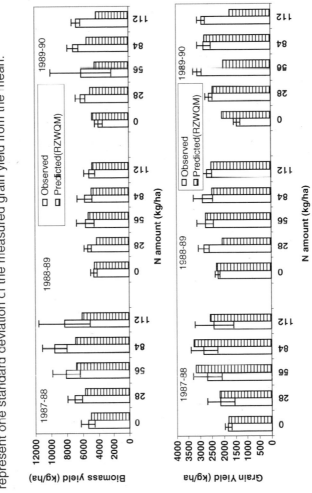

FIGURE 7. Predicted soybean yields via RZWQM and a locally derived regression equation. Irrigation was applied with a gradient line-source system. 1985 irrigation (cm): (1) 0.28, (2) 3.38, (3) 8.86, and (4) 12.92. 1986 irrigation (cm): (1) 1.15, (2) 7.22, (3) 17.11, and (4) 24.98.

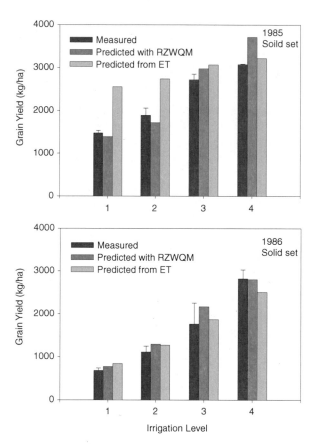

farmers) with computer-based simulation modeling to engage small-holder farmers in Zimbabwe on issues of soil fertility management. One of their applications involved the simulation of different levels of manure application to answer a local farmer's question about the value of applying manure. Through a combination of discussions with farmers and simulations with the APSIM cropping systems model, they demonstrated to the local farmers that the addition of low quality manure (Figure 11 b, C:N = 35:1) would have a detrimental effect on maize yield

FIGURE 8. Water and nitrogen interaction effects on maize grain yield using DSSAT models. (Singh et al., 2002. Copyright 2002 from Agricultural System Models in Field Research and Technology Transfer by L.R. Ahuja, L. Ma, and T.A. Howell (Eds.). Reproduced by permission of Routledge/Taylor & Francis Group, LLC.)

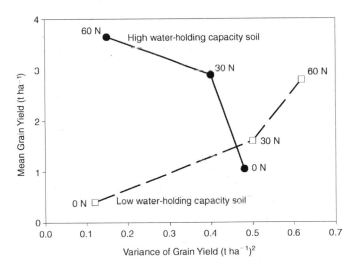

(because of N immobilization) but that the addition of higher-quality manure could increase maize yields (Figure 11 c and d).

A major concern in crop production is the sustainability of existing cropping systems and possibilities for improvement through introduction of alternate crops in a rotation. In a study of dry land cropping systems in eastern Colorado, Andales et al. (2003) used both GPFARM and experimental field data to show that cropping intensification more effectively utilized available soil moisture and increased overall system productivity compared with the prevalent wheat-fallow rotation (Figure 12).

In a simulation exercise using the Decision Support System for Agrotechnology Transfer (DSSAT) models in Brazil, Bowen et al. (1998) found that a continuous maize-fallow system with no inputs of fertilizer exhibited a gradual decline in maize yields across 50 years. On the other hand, a green manure-maize-fallow system was shown to maintain yields during the same period.

FIGURE 9. The responses of yearly average N uptake, maize silage yield, and nitrate-N leached to irrigation according to degree of root zone water depletion, as simulated by RZWQM.

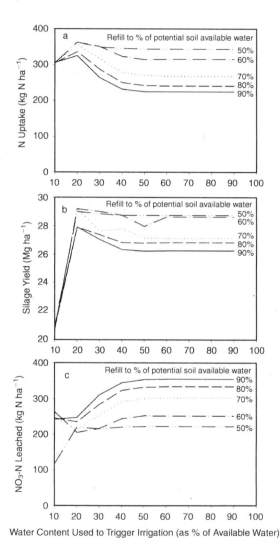

FIGURE 10. Simulated responses of N uptake, maize silage yield, and nitrate-N leaching to the timing of N application. Manure was NOT applied and water was applied 4-6 times a year (20 cm/event).

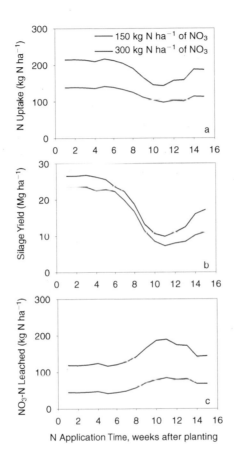

The International Rice Research Institute is using rice crop models to help design morphological traits of the new plant types for yield and weed competition (Dingkuhn et al., 1991). Similarly the Pioneer Hi-Bred Company is exploring the use of models for developing new maize hybrids (personal communication).

Some examples of the commercial use of models to guide tactical management are (Matthews and Stephens, 2002):

1. SIRATAC for insecticide applications in cotton in Australia, 1981-89.
2. EMPIRE for control of diseases and pests in winter wheat in Europe, 1980s.
3. PUTU for irrigation scheduling in S. Africa, 1980s.
4. CROPWAT for developing irrigation scheduling guidelines by FAO, 1980-90.

The models are also being used in teaching in several different countries.

Water Quality

An important requirement for the sustainability of current agricultural systems is the mitigation of adverse environmental effects. Intensive crop production has been recognized as a significant non-point source of water contaminants. A major concern is the movement of nitrate (NO_3-N), phosphorus, and agricultural chemicals (e.g., pesticides, herbicides) from agricultural fields into surface and ground water bodies. Agricultural system models that have the capability to simulate transformations and movement of agricultural chemicals have been used to assess the interactions between water quality and crop production management.

The Root Zone Water Quality Model (RZWQM) has been applied to various investigations of nitrate movement in the root zone and subsurface drains. Singh (1994) used the model to predict NO_3-N concentrations in the soil profile under different tillage practices in Iowa (Figures 13 and 14). He showed that the model could accurately differentiate among four alternative tillage practices and that NO_3-N concentration in the soil profile was a function of the degree of soil disturbance by tillage (e.g., no-till exhibited the least NO_3-N concentrations). In the same study, the NO_3-N concentrations in tile flow (sub-surface drainage) were also simulated (Figure 15).

Chemical transport in agricultural soils is greatly influenced by soil texture and structure. For example, the presence of surface aggregates and macropores (e.g., worm holes, channels left by decayed roots, cracks in drying soil) can result in preferential flow (i.e., by-passing of the soil matrix) and transport of pollutants to groundwater. The RZWQM was successfully used to simulate the flow of bromide in soil columns of varying combinations of surface aggregation and presence of macropores (Ahuja et al., 1995). The concentrations of

FIGURE 11. (a) Baseline simulation, using climate data for Tsholotsho for 11 years (1991-2001), was for maize cultivar sc501 grown on *ipane* soil with no applications of manure or inorganic fertilizer; and the changes in maize yields (bags acre^{-1}); (b) for the application of low quality manure; (c) for the application of high quality manure; and (d) for the application of high quality manure concentrated on a smaller area (0.5 acre). (Carberry et al., 2004. Reproduced by permission of ACIAR.)

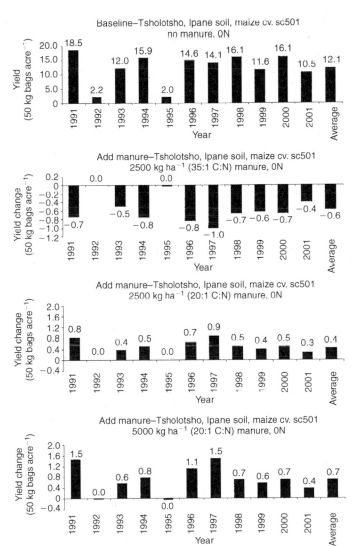

FIGURE 12. Annualized yields at Sterling, CO (1989-1993) for different crop rotations (WF = wheat-fallow, WCF = wheat-corn-fallow, WCMF = wheat-corn-millet-fallow), observed and simulated by GPFARM.

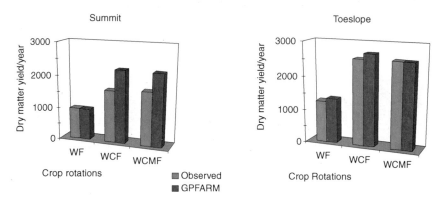

FIGURE 13. Observed (points) vs. RZWQM simulated (lines) NO$_3$-N concentration for soil profile for day 150 in 1990. CP-Chisel Plow, MP-Moldboard Plow. (Modified from Singh, 1994.)

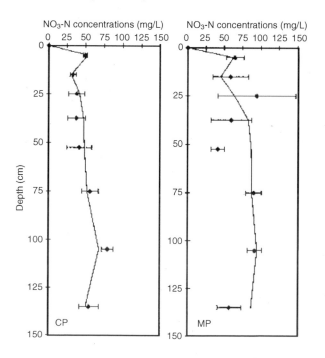

FIGURE 14. Observed (points) vs. RZWQM simulated (lines) NO$_3$-N concentration for soil profile for day 150 in 1990. NT-No Tillage, RT-Reduced Tillage. (Modified from Singh, 1994.)

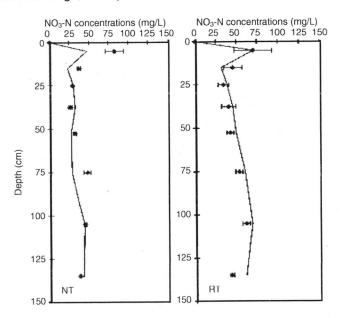

FIGURE 15. Water and nitrate movement into tile drains (Nashua, IA). Simulations were done with RZWQM. (Modified from Singh, 1994.)

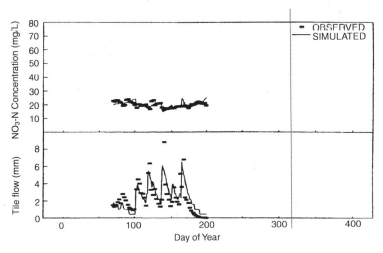

bromide, which is a surrogate for non-absorbed chemicals such as NO_3-N, were adequately simulated in the soil columns (Figures 16 and 17). This provided some evidence that RZWQM can be used to predict transport of non-absorbed chemicals in soils exhibiting preferential flow.

In a field application of the RZWQM in Georgia, Ma et al. (1995) simulated atrazine (herbicide) transport in runoff. A strong correlation between simulated and measured atrazine in runoff (Figure 18) sug-

FIGURE 16. Bromide concentrations in soil water observed (solid curve) and simulated with RZWQM (dashed curve) with no surface aggregates.

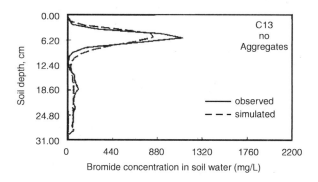

FIGURE 17. Bromide concentration in soil water as influenced by macropores and a 1 cm layer of surface aggregates observed (solid curve) and simulated with RZWQM (dashed curve).

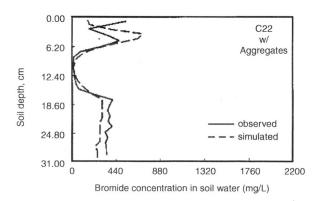

gested that the model effectively simulated movement of atrazine from the field.

Climate Change Effects on Crop Production

At the close of the 20th Century, widespread concern grew about climate change brought about by anthropogenic pollution. In the agricultural sector, the main concern is the effect of climate change (e.g., increased atmospheric CO_2, elevated air temperature) on crop production. Agricultural system models have been used to investigate the possible impacts of climate change on yield. Practically, system modeling is the only feasible approach to the study of this global phenomenon as it is impossible to completely understand the interactions between climate change and crop production based on limited plot-scale experiments or controlled-environment studies. With system models, investigators can simulate crop production under various scenarios of climate change.

The GOSSYM cotton model was used in a 30-year simulation study and showed that increased CO_2 had a positive effect on cotton production in Mississippi (Figure 19; Reddy et al., 2002). Various climate scenarios

FIGURE 18. Relationship between measured and simulated atrazine in runoff, Watkinsville, GA, 1973-1975.

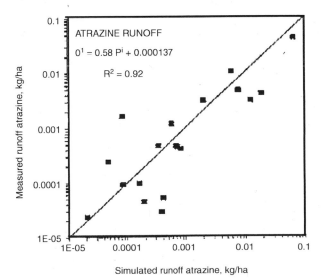

Simulated runoff atrazine, kg/ha

FIGURE 19. Simulated cotton lint yield response to carbon dioxide enrichment. (Reddy et al., 2002. Copyright 2002 from Agricultural System Models in Field Research and Technology Transfer by L.R. Ahuja, L. Ma, and T.A. Howell (Eds.). Reproduced by permission of Routledge/Taylor & Francis Group, LLC.)

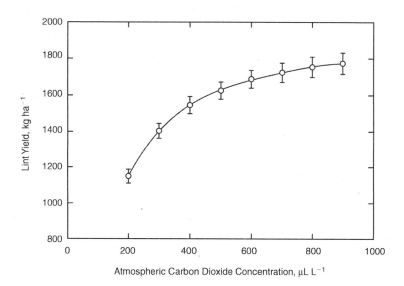

Atmospheric Carbon Dioxide Concentration, μL L⁻¹

were investigated based on possible combinations of CO_2 concentration and weather patterns (Figure 20; Reddy et al., 2002).

In a study of climate change effects on rice yields in India, Aggarwal and Mall (2002) looked at the interactions among uncertainties in climate change scenarios, crop models, and nitrogen management (Figure 21). There was considerable difference in the impact of climate change on rice yields calculated by the ORYZA1N and CERES-Rice crop models (see differences between lines 1 and 2; lines 3 and 4 in Figure 21). This example shows the potential for arriving at different conclusions, mainly due to differences in assumptions built into different crop models.

A good example of a large-scale application of crop modeling in climate change studies is the investigation by Matthews et al. (1995) funded by the U.S. Environmental Protection Agency. The ORYZA1 model was used to predict changes in rice production for the major rice-producing countries in Asia under three general circulation model (GCM) scenarios. In general, an increase in CO_2 level was found to increase rice yields, whereas yields were reduced with increases in temperature.

FIGURE 20. Simulated cotton yields for different years with varying weather patterns. (Reddy et al., 2002. Copyright 2002 from Agricultural System Models in Field Research and Technology Transfer by L.R. Ahuja, L. Ma, and T.A. Howell (Eds.). Reproduced by permission of Routledge/Taylor & Francis Group, LLC.)

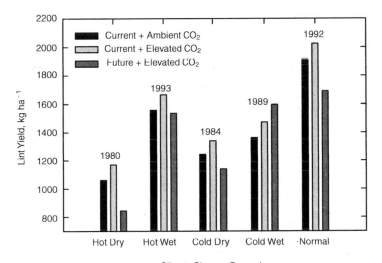

Not only does climate change affect crop production, but also contributes to further climate change via production of greenhouse gases. Methane (CH_4) is a greenhouse gas that is emitted from rice fields in significant amounts because of the anaerobic soil conditions under which rice is grown. Matthews et al. (2000) used the MERES (Methane Emissions from Rice EcoSystems) model to upscale experimental field data of CH_4 emissions to the national level and to evaluate potential mitigation strategies. The model predictions showed that field drainage to minimize anaerobic conditions could potentially decrease CH_4 emissions by an average of 13% across five countries, viz., China, India, Indonesia, Philippines, and Thailand.

COLLABORATIONS FOR FURTHER DEVELOPMENTS

The collective experiences of model developers and users show that, even though they are not perfect, the agricultural system models can be

FIGURE 21. Response of irrigated rice yields (% change in yield in climate change over the control yield) in northern India to different levels of N availability for the Intergovernmental Panel on Climate Change's (IPCC) optimistic and pessimistic scenarios of climate change for 2010 and 2070. The simulations were done using two crop models–CERES-rice and ORYZA1N. The difference between lines 1 and 2 and between lines 3 and 4 refers to uncertainties in impact assessment due to climate change scenarios as simulated by ORYZA1N and CERES-rice, respectively. The difference between the top and bottom lines in each figure refers to the total uncertainties due to crop models and climate change scenarios. (Aggarwal and Mall, 2002. Reproduced by permission of Springer Science and Business Media.)

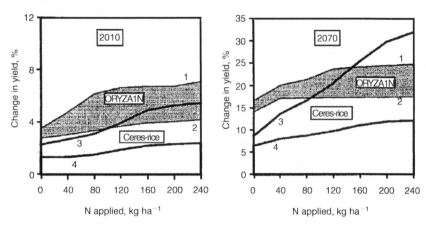

very useful in field research, technology transfer, and management decision-making as demonstrated in this paper. These experiences also show a number of problems or issues that should be addressed to improve the models and their applications. The most important issues are:

- System models need to be more thoroughly tested and validated for science defensibility under a variety of soil, climate, and management conditions, with experimental data of high resolution in time and space.
- There is a need to build comprehensive and common shared experimental databases based on existing standard experimental protocols, and relate measured values to modeling variables, so that conceptual model parameters can be experimentally verified.
- There is a need for better methods of determining parameters for different spatial and temporal scales, and for aggregating simulation results from plots to fields and larger scales.

- There is a need for better communication and coordination among model developers in the areas of model development, model parameterization, and model evaluation.
- There is a need for better collaboration between model developers and field scientists for appropriate experimental data collection and for evaluation and application of models. Many times the involvement of field scientists in modeling exercises is limited to providing experimental data for model testing. Instead, field scientists should be involved in model development from the beginning.
- There is an urgent need for filling the most important knowledge gaps: agricultural management effects on soil-plant-atmosphere properties and processes; plant response to water, nutrients, and temperature stresses; and effects of natural hazards like hail, frost, insects, and diseases.
- And finally we need to improve upon the methods and structure of model building so that: (1) the models are modular, with each model component (module) clearly defined, documented, and assigned a degree of uncertainty; (2) each model component can be independently tested and improved, and can be easily substituted; (3) the whole world community can contribute to developing, testing, and improving components; (4) the components may vary with the scale of application; (5) hierarchical parameter estimation from varying degrees of input information is a component of the model; (6) the assembled models of the system are kept compact and easy to use by customizing them to agro-ecosystem regions; (7) a user-friendly interface is provided for easy input of data and output of results; and (8) a well-illustrated user manual is provided to illustrate a step-by-step procedure for running the model and some examples of model application that demonstrate the benefits of using the model as well as the uncertainty in results.

To address the last issue stated above, a cross-agency (USDA-ARS, NRCS, USGS) project is underway to develop a modular modeling computer framework that will consist of a library of alternate modules (or subroutines) for different sub-processes of science, associated databases, and the logic to facilitate the assembly of appropriate modules into a modeling package (David et al., 2002; Ahuja et al., 2005). The modeling package can be tailored or customized to a problem, data constraints, and scale of application. The framework will: (1) enable the use of best science for all components of a model; (2) allow quick updates or replacement of science or database modules as new knowledge becomes available; (3) eliminate duplication of work by modelers; (4) provide a common platform and standards for development and implementation; (5) serve as a reference and coordination mechanism for future research and development; and (6) make collaboration much easier among modelers by sharing science modules/components and experimental/simulated databases, so that specialties of each individual modeling group can be maximally utilized.

In the future, model developers need to work together among themselves to address the seven problem areas described above, and then train and work with field scientists to improve model visibility and applicability in solving real-world problems. Also, there is a need to better document system models and simulated processes so that field scientists will be able to understand these processes without too much difficulty. We also need to document good case studies on model applications to serve as guides for field users. Any improvements to an existing model could be checked against these documented cases to see if these improvements are applicable to all situations. Since most field data are not collected for the purpose of evaluating with a system model, some good system-oriented experiments may be needed. International efforts are needed to coordinate system modeling and to encourage model developers and field scientists to work on identified knowledge gaps and research priorities. The above actions will prepare the models for their important roles in the 21st Century, and take the agricultural research and technology to the next higher plateau.

CONCLUSIONS

This article describes the tremendous potential benefits of the agricultural system models for field research and as management guides. It provides some examples of model applications in research and management from the worldwide literature. It also brings out the limitations and knowledge gaps in the current models where further improvements must be made.

Where do we go from here? In order to gradually realize the potential benefits of the models and further improve the models, we believe the most important first step is to fully integrate system models with field research. Modelers must work collaboratively with field scientists in various areas from several different locations. Even with current shortcomings of the models, the models can be very useful in synthesizing and quantifying experimental results across different climates, soils, and agricultural systems at different locations. This synthesis will enhance understanding of the diversity of results and point toward different management practices. For example, system models should be used to synthesize results for different crop rotations, water quality, and greenhouse gas emissions across a region or a country. The models should also be used to evaluate sustainability of current agricultural systems over longer periods of time, with current climate and with projected climate changes. This will lead to information for devising management strategies to mitigate adverse effects of climate change. The major new findings from the regional or national evaluations and from long-term system analysis could be provided to the end users and policy makers in simple spreadsheet and graphical forms.

REFERENCES

Aggarwal, P. K., Kalra, N., Bandyopadhyay, S. K., and Selvarajan, S. 1995. A systems approach to analyze production options for wheat in India. In: Bouma, J., Kuyvenhoven, A., Bouman, B. A. M., Luten, J. C., and Zandstra, H. G. (Eds.). Eco-Regional Approaches for Sustainable Land Use and Food Production. Systems Approaches for Sustainable Agricultural development. Kluwer, Dordrecht, pp. 167-186.

Aggarwal, P. K. and Mall, R. K. 2002. Climate change and rice yields in diverse agro-environments of India. II. Effect of uncertainties in scenarios and crop models on impact assessment. Climatic Change 52:331-343.

Ahuja, L. R., Johnsen, K. E., and Heathman, G. C. 1995. Macropore transport of a surface-applied bromide tracer: Model evaluation and refinement. Soil Sci. Soc. Am. J. 59:1234-1241.

Ahuja, L. R. and Ma, L. 2002. Computer Modeling of Water Management. Encyclopedia of Soil Science. Marcel Dekker Inc., 270 Madison Avenue, New York, NY 10016, pp. 218-222.

Ahuja, L. R., Ma, L., and Howell, T. A. (Eds.). 2002. Agricultural System Models in Field Research and Technology Transfer. CRC Press, Boca Raton, FL, 357 pp.

Ahuja, L. R., Rojas, K. W., Hanson, J. D., Shaffer, M. J., and Ma, L. (Eds.). 2000. Root Zone Water Quality Model. Water Resources Publications, LLC, Highlands Ranch, CO, 372 pp.

Ahuja, L. R., Ascough, II, J. C., and David, O. 2005. Developing natural resource models using the Object Modeling System: Feasibility and challenges. Advances in Geosciences 4:29-36.

Andales, A. A., Ahuja, L. R., and Peterson, G. A. 2003. Evaluation of GPFARM for dryland cropping systems in Eastern Colorado. Agronomy J. 95:1510-1524.

Ascough II, J. C., Hanson, J. D., Shaffer, M. J., Buchleiter, G. W., Bartling, P. N., Vandenberg, B. C., Edmunds, D. A., Wiles, L., McMaster, G. S., and Ahuja, L. R. 1995. The GPFARM decision support system for whole farm/ranch management. Proc. Workshop on Computer Applications in Water Management; Great Plains, Agr. Council Public. 154. Fort Collins, CO, pp. 53-56.

Bowen, W. T., Thornton, P. K., and Hoogenboom, G. 1998. The simulation of cropping sequences using DSSAT. In: Tsuji, G. Y., Hoogenboom, G., and Thornton, P. K. (Eds.). Understanding Options for Agricultural Production. Systems Approaches for Sustainable Agricultural Development. Kluwer, Dordrecht, pp. 313-327.

Carberry, P. S., Gladwin, C., and Twomlow, S. 2004. Linking simulation modelling to participatory research in smallholder farming systems. In: Delve, R. J. and Probert, M. E. (Eds.). Modelling Nutrient Management in Tropical Cropping Systems. ACIAR Proceedings no. 114. Australian Centre for International Agricultural Research, Canberra, pp. 32-46.

David, O., Markstrom, S. L., Rojas, K. W., Ahuja, L. R., and Schneider, I. W. 2002. The object modeling system. In: Ahuja, L. R., Ma, L. and Howell, T. A. (Eds.). Agricultural System Models in Field Research and Technology Transfer. Lewis Publishers, Boca Raton, FL pp. 317-330.

Dimes, J., Twomlow, S., and Carberry, P. 2003. Application of APSIM in smallholder farming systems in the semi-arid tropics. In: Struif Bontkes, T. E. and Wopereis, M. C. S. (Eds.). Decision Support tools for Smallholder Agriculture in Sub-Saharan Africa: A Practical Guide. International Center for Soil Fertility and Agricultural Development (IFDC), Muscle Shoals, AL and ACP-EU Technical Centre for Agricultural and Rural Cooperation (CTA), Wageningen, The Netherlands, pp. 85-99.

Dingkuhn, M., Penning de Vries, F. W. T., De Datta, S. K., and van Laar, H. H. 1991. Concepts for a new plant type for direct seed flooded tropical rice. In: Direct Seeded Flooded Rice in the Tropics: Selected Papers from the International Rice Research Conference. International Rice Research Institute, Los Baños, The Philippines, pp. 17-38.

Dzotsi, K., Agboh-Noameshie, A., Struif Bontkes, T. E., Singh, U., and Dejean, P. 2003. Using DSSAT to derive optimum combinations of cultivar and sowing date for maize in Southern Togo. In: Struif Bontkes, T. E. and Wopereis, M. C. S. (Eds.). Decision Support tools for Smallholder Agriculture in Sub-Saharan Africa: A Practical Guide. International Center for Soil Fertility and Agricultural Development (IFDC), Muscle Shoals, AL and ACP-EU Technical Centre for Agricultural and Rural Cooperation (CTA), Wageningen, The Netherlands, pp. 100-113.

Heilman, P., Hatfield, J. L., Rojas, K. W., Ma, L., Huddleston, J., Ahuja, L. R., and Adkins, M. 2002. How good is good enough? What information is needed for agriculture and how can it be provided most affordably? J. Soil & Water Conserv. 57:98-105.

Ma, L., Ahuja, L. R., Ascough II, J. C., Shaffer, M. J., Rojas, K. W., Malone, R. W., and Cameira, M. R. 2000. Integrating system modeling with field research in agriculture: Applications of Root Zone Water Quality Model (RZWQM). Advances in Agronomy 71: 233-292.

Ma, Q. L., Ahuja, L. R., Rojas, K. W., Ferreira, V. F., and DeCoursey, D. G. 1995. Measured and RZWQM predicted atrazine dissipation and movement in a field soil. Transactions of the American Society of Agricultural Engineers 38: 471-479.

Matthews, R. B., Kropff, M. J., Horie, T., and Bachelet, D. 1995. Simulating the impact of climate change on rice production in Asia and evaluating options for adaptation. Ag. Systems 54:399-425.

Matthews, R. B. and Stephens, W. 1998. The role of photoperiod in regulating seasonal yield variation in tea (*Camellia sinensis* L.). Experimental Agriculture 34: 323-340.

Matthews, R. B. and Stephens, W. (Eds.). 2002. Crop-Soil Simulation Models: Applications in Developing Countries. CABI Publishing, UK, 277 pp.

Matthews, R. B., Wassmann, R., Knox, J. W., and Buendia, L. 2000. Using a crop/soil simulation model and GIS techniques to assess methane emissions from rice fields in Asia. IV. Upscaling of crop management scenarios to national levels. Nutrient Cycling in Agroecosystems 58:201-217.

Reddy, K. R., Kakani, V. G., McKinion, J. M., and Baker, D. N. 2002. Applications of a cotton simulation model, GOSSYM, for Crop Management, Economic, and Policy Decisions. In: Ahuja, L., Ma, L. and Howell, T. A. (Eds.). Agricultural System Models in Field Research and Technology Transfer. Lewis Publishers, Boca Raton, FL, pp. 33-54.

Rojas, K. W., Heilman, P., Huddleson, J., Ma, L., Ahuja, L. R., Hatfield, J. L., and Kasireddy, S. 2000. An integrated research information and decision support system for conservation planning and management. Agronomy Abstracts. p. 419.

Saseendran S. A., Nielson, D. C., Ma, L., Ahuja, L. R., and Halvorson, A. D. 2004. Modeling nitrogen management effects on winter wheat production using RZWQM and CERES-wheat. Agronomy Journal 96:615-630.

Singh, P. 1994. Modification of RZWQM to simulate the tillage effects on subsurface drain flows and NO_3-N movement. Unpublished PhD dissertation. Iowa State University, Ames, IA, 175 pp.

Singh, U., Wilkens, P. W., Baethgen, W. E., and Bontkes, T. S. 2002. Decision support tools for improved resource management and agricultural sustainability. In: Ahuja, L. R., Ma, L. and Howell, T. A. (Eds.). Agricultural System Models in Field Research and Technology Transfer. Lewis Publishers, Boca Raton, FL, pp. 91-117.

Struif Bontkes, T. E. and Wopereis, M. C. S. (Eds.). 2003. Decision Support Tools for Smallholder Agriculture in Sub-Saharan Africa: A Practical Guide. International Center for Soil Fertility and Agricultural Development (IFDC), Muscle Shoals, AL and ACP-EU Technical Centre for Agricultural and Rural Cooperation (CTA), Wageningen, The Netherlands, 194 pp.

Wopereis, M. C. S., Haefele, S. M., Dingkuhn, M., and Sow, A. 2003. Decision support tools for irrigated rice-based systems in the Sahel. In: Struif Bontkes, T. E. and M. C. S. Wopereis (Eds.). Decision Support Tools for Smallholder Agriculture in Sub-Saharan Africa: A Practical Guide. International Center for Soil Fertility and Agricultural Development (IFDC), Muscle Shoals, AL and ACP-EU Technical Centre for Agricultural and Rural Cooperation (CTA), Wageningen, The Netherlands, pp. 114-126.

doi:10.1300/J411v19n01_04

Sustainability Considerations
in Wheat Improvement and Production

S. Rajaram

K. D. Sayre

J. Diekmann

R. Gupta

W. Erskine

SUMMARY. This article describes the global wheat mega-environments and consequently need for different types of wheat germplasm. The breeding programs worldwide have targeted yield potential gains as one of their major objectives. The yield gains have been variable, but consistently increasing until the end of the last century. However, in general, there has been a large benefit (estimated between 2 to 6 billion US\$ based on year 2000 parity) to the agricultural economy in developing countries due to international and national partnerships in wheat

---

S. Rajaram, J. Diekmann, and W. Erskine are affiliated with ICARDA, P.O. Box 5466, Aleppo, Syria (E-mail: s.rajaram@cgiar.org).

K. D. Sayre is affiliated with CIMMYT, Apartado 6 641 Colonia Juarez, 06600 Mexico DF, Mexico.

R. Gupta is affiliated with CIMMYT, Rice-Wheat Consortium for The Indo-Gangetic Plains (RWC), CG Central Block, National Agricultural Science Center (NASC) Complex, D. P. Shastri Marg, Pusa Campus, New Delhi 110012, India.

Presented to Sustainability of Agriculture, Environment & Food Security in Developing Countries, November 7, 2005, Salt Lake City, USA.

[Haworth co-indexing entry note]: "Sustainability Considerations in Wheat Improvement and Production." Rajaram, S. et al. Co-published simultaneously in *Journal of Crop Improvement* (Haworth Food & Agricultural Products Press, an imprint of The Haworth Press, Inc.) Vol. 19, No. 1/2 (#37/38), 2007, pp. 105-123; and: *Agricultural and Environmental Sustainability: Considerations for the Future* (ed: Manjit S. Kang) Haworth Food & Agricultural Products Press, an imprint of The Haworth Press, Inc., 2007, pp. 105-123. Single or multiple copies of this article are available for a fee from The Haworth Document Delivery Service [1-800-HAWORTH, 9:00 a.m. - 5:00 p.m. (EST). E-mail address: docdelivery@haworthpress.com].

breeding. The production and productivity of wheat in India has begun to show stagnation, primarily due to natural resource base decline. The research results indicate that this can be reversed by practicing zero tillage and timely planting in the Gangetic Plains. The dryland wheat based agriculture is also declining, especially in WANA (West Asia and North Africa). However, the results pertaining to minimum tillage, timely sowing and lentil/vetch rotation with wheat have shown promise to provide high productivity of wheat (up to 4 ton/ha) when the temperature extremes (freezing and very hot) are not common. This article also provides evidence of sustainability and profitability of irrigated agriculture in Sonora, Mexico, provided farmers practice zero tillage, residue management, and raised-bed planting system. doi:10.1300/J411v19n01_05 *[Article copies available for a fee from The Haworth Document Delivery Service: 1-800-HAWORTH. E-mail address: <docdelivery@haworthpress.com> Website: <http://www.HaworthPress.com> © 2007 by The Haworth Press, Inc. All rights reserved.]*

KEYWORDS. Mega-environment, wheat germplasm, wheat breeding, zero tillage, sustainable wheat production, drought-tolerant cultivars, yield gains

Bread wheat (*Triticum aestivum* L.) is the most widely grown and consumed food crop. It is the staple food of nearly 35% of the world population and demand for wheat grows faster than for any other major crop. The forecasted global demand for wheat in 2020 varies between 840 and 1050 million tons (Kronstad, 1998; Rosegrant et al., 1995). To reach this target, global production will need to increase 1.6 to 2.6% annually from the present production level of 560 million tons. Increases in realized grain yield have provided about 90% of the growth in world cereal production since 1950 (Mitchell et al., 1997) and by the end of the first decade of the 21st Century, most of the increase needed in world food production must come from higher absolute yields (Ruttan, 1993). For wheat, the global average grain yield must increase from the current 2.7 to 3.8 t ha^{-1} (Figure 1). The future increases in food productivity will require substantial research and development investments to improve the profitability of wheat production systems through enhancing input-use efficiency along with other important crops like rice, maize, millets, and tubers. A global targeting of wheat average yield of 3.8 t ha^{-1} by 2020 is a necessary step towards meeting the UN millennium goal.

FIGURE 1. Wheat Yield Worldwide Over Time

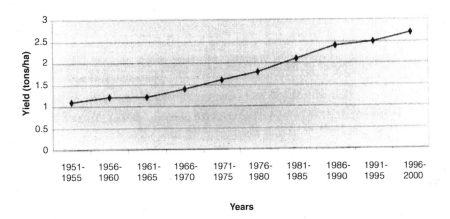

WHEAT PRODUCTION MEGA-ENVIRONMENTS

Wheat as species is broadly classified into spring habit that is lacking vernalization and winter and facultative (W/F) habits that require vernalization to a variable degree. Wheat varieties both from spring and W/F habits can differ further in their requirement of photoperiod. Some varieties are photoperiod sensitive whereas others are neutral. Similarly, varieties differ widely in their response to high and low temperatures, moisture regimes, disease and insect resistance, pH and quality traits. The cultivated types are predominantly bread wheat (90% of cultivated areas), and durum wheats (8%). A very small percent of area is still planted to Dicoccum and Eionkorn, especially in India and the Middle East where wheat originated.

International wheat breeding programs, especially those conducted by CIMMYT and ICARDA, categorize environmental factors into homogeneous groups and design experiments that permit production of genotypes that are adapted, encompassing various useful traits. This has permitted the CGIAR Centers to produce germplasm whose adaptation is international, releasing internal public goods globally.

Rajaram et al. (1994) described the conceptual framework of mega-environments (ME) for international wheat breeding. We have presented here a simplified version of ME from the original 12 to 5 (Table 1). These MEs are inter-continental, not necessarily contiguous, and have factors of relative homogeneity for moisture regimes, and photoperiod and ver-

TABLE 1. Global Wheat Mega-Environments (200 million hectares)

ME		Areas Million of Hectares
1	Irrigated spring wheat (Low latitudes) Punjab, India; Sonora, Mexico	32 million
2	Low rainfall semiarid spring wheat (Low latitudes) Syria, Morocco	60 million
3	Spring sown spring wheat (High latitudes) Northern Kazakhstan	8 million
4	Irrigated W/F wheat High Plateau of Turkey	100 million
5	Semiarid low rainfall W/F wheat Marageh, Iran	
	Total	200 million

nalization requirements. Broadly defined, these MEs are: (1) Irrigated Spring Wheat; (2) Low Rainfall Semi-Arid Spring Wheat; (3) Spring-Sown Spring Wheat; (4) Irrigated W/F Wheat; and (5) Semi-Arid Low Rainfall W/F Wheat.

INTERNATIONAL WHEAT BREEDING AND YIELD GAINS

The genetic basis of the Green Revolution in wheat production involved the introduction into wheat of a few genes with major effect on plant height and the elimination of photoperiod response. These radical changes in plant architecture and phenology, accompanied by improved disease resistance, greatly enhanced the yield potential of released varieties in many parts of the world. Since the introduction of semi-dwarf wheat cultivars, wheat yields have continued to improve at an average rate of 1% per year (Tables 2 and 3). Note, however, that most of the studies in relation to yield potential cover the period until the early 1990s. It is not clear whether this trend has been maintained since then.

Irrespective of whether yield gains are maintained, international wheat breeding conducted by CGIAR Centers (CIMMYT and ICARDA) and a vast network of National Agricultural Research Systems (NARS) has provided substantial benefits for the livelihood of poor farmers in many developing countries. In Table 4, these benefits are listed based on the assumption of yield gain of 0.15, 0.25, 0.35, and 0.45 t ha^{-1}. The additional amount of wheat produced in developing countries that is attributed to international wheat breeding research is estimated to range from 13 million tons per year under the most conservative scenario to 41 mil-

TABLE 2. Rates of Genetic Gains in Bread Wheat Grain Yield in Developing Countries

Environment/Location	Period	Rate of gain (%/yr)	Source
Sonora, Mexico	1962-83	1.1	Waddington et al. (1986)
	1962-88	0.9	Sayre et al. (1997)
India	1967-79	1.2	Kulshrestha and Jain (1982)
	1989-99	1.9	Nagarajan (2002)
Zimbabwe	1967-85	1.0	Mashiringwani (1987)
Argentina	1966-89	1.9	Byerlee and Moya (1993)
Rio Grande do Sul, Brazil	1970-90	3.6	Tomasini (2002)
South Africa	1930-90	1.4	Van Lill and Purchase (1995)

TABLE 3. Rates of Genetic Gains in Bread Wheat Grain Yield in Developed Countries

Environment/Location	Period	Rate of gain (%/yr)	Source
New South Wales, Australia	1956-84	0.9	Anthony and Brennan (1987)
Kansas, USA	1976-87	1.2	Cox et al. (1988)
UK	1947-77	1.5	Silvey (1978)

TABLE 4. Global Benefits from International Wheat Breeding Research

Assumed grain yield from MV's (t/ha)	Additional annual production (million t)	Value of additional annual production (US$ billion 2002)
0.15	13.7	2.0
0.25	22.8	3.4
0.35	31.9	4.8
0.45	41.0	6.1

Modern varieties (MVs) planted on 91.1 million hectares; assumed price of wheat is US$ 150/t (2002 dollars)
Adapted from Lantican et al. (2005)

lion tons under the most liberal scenario. Converting these physical quantities into economic terms, the total value of additional wheat grain production in developing countries that is attributable to international wheat improvement research ranges from US$ 2.0 billion to US$ 6.1 billion per year (Lantican et al., 2005).

ISSUES OF SUSTAINABLE WHEAT PRODUCTION IN INDIA

As indicated in Table 5, it appears that the Indian wheat productivity has plateaued in the last four years (2000-2003 period). The average productivity for this period is about 27 Q ha^{-1}. The Indian Coordinated Wheat Program has done an excellent job of identifying some of the most elite germplasm from international centers. Consequently, the productivity increases in part occurred from the average of 25 Q ha^{-1} to 27 Q ha^{-1}. Some of the elite germplasm is listed in Table 6. The variety PBW343 derived from the cross 'Attila', has shown a 10% yield advantage in many experiments in Northwest India over the previously widely grown variety HD2328. The economic impact of the variety alone is substantial and it is now grown on 5.73 million hectares (Lantican et al., 2005).

The wheat projection statistics for India for 2020 are presented in Table 7. The wheat demand for 2020 is projected to be 100 million tons. The current production is about 70 million tons (Table 5). It is expected that India would have to import at least between 6 to 7 million tons of wheat with an approximate value of 1.2 to 1.4 billion dollars despite continuing maintenance production.

Results of an experiment conducted under Rice-Wheat System by R. Gupta and colleagues indicate that yield plateau can be raised by adopting better agronomic practices, such as conservation agriculture with zero tillage in the rice-wheat system. The data from Bihar, India has

TABLE 5. Area, Production, and Productivity of Wheat in India (1965-2003)

Year	Area (MHa)	Production (MT)	Yield (Q/Ha)
1965	13.4	12.3	9.1
1970	16.6	20.1	12.1
1975	18.0	24.1	13.3
1980	22.2	31.8	14.3
1985	23.6	44.1	18.7
1990	23.5	49.8	21.2
1995	25.7	65.8	25.6
2000	27.6	76.4	27.8
2001	25.7	61.7	27.1
2002	25.7	71.8	27.7
2003	25.2	69.3	27.5

Adapted from Wheat Project Directorate Karnal, 2005

TABLE 6. Hallmark Wheat Varieties

Cross	Year cross made	Year of release	Area planted, 1997 (m.ha)	Area planted, 2002 (m.ha)
Veery	1974	1981	3.35	0.60
Bobwhite	1974	1983	1.60	0.98
Kauz	1980	1988	1.90	1.73
Attila	1984	1995	1.00	5.73

Adapted from Lantican et al. (2005)

TABLE 7. Wheat Projection Statistics for India from 1997-2020

Area	Year	
	1997	2020
Area (000 ha)	25902	27776
Yield (kg/ha)	2552	3383
Production (000 MT)	66105	93955
Demand (000 MT)	66857	100686
Food (000 MT)	58465	87629
Feed (000 MT)	789	1606
Net Trade (000 MT)	−761	−6731

Adapted from IFPRI, Impact Global Food Production

consistently shown productivity advantage of 247 kg ha^{-1} over conventional practices followed now. The research trials conducted since 1997-1998 to 2004-2005 suggest this advantage is consistent (Figures 2 and 3). The zero tillage can be recommended for the entire rice-wheat system, which is now widely considered to be deteriorating due to various factors, but primarily due to late planting, unbalanced soil structure, and micronutrient deficiency. If small machineries are made available to farmers, the zero-tillage system can raise total production in the order of 2.5 million tons in the Gangetic Plains.

SUSTAINABLE MANAGEMENT OF WHEAT UNDER IRRIGATION IN SONORA, MEXICO (PERMANENT BEDS)

A shift from conventional tillage to a reduced or zero-till seeding system with residue retention may require several crop cycles before po-

FIGURE 2. Effect of Zero Tillage Over Conventional Tillage in Wheat Productivity in India

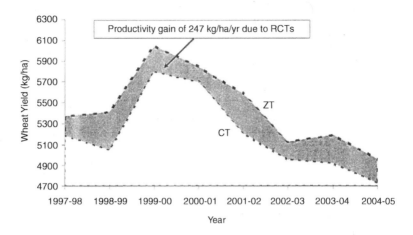

FIGURE 3. Effect of Zero Tillage on Wheat Productivity in Bihar, India

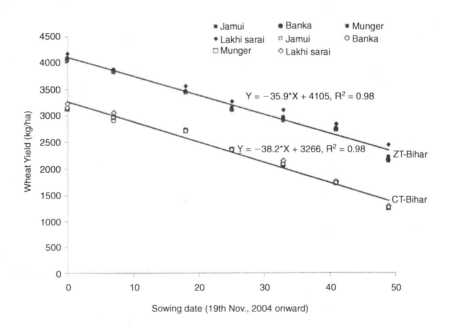

tential advantages/disadvantages begin to become apparent (Blevins et al., 1984) and the results from this long-term trial confirm this observation.

Results from various tillage systems conducted across 12 wheat crop cycles with 300 kg N ha^{-1} in Sonora, Mexico, are presented in Figure 4. Only small yet significant yield differences between the tillage/residue managements occurred from 1993 to 1997. However, beginning with the 1998 wheat crop, large and significant differences between the tillage/residue management options occurred even though year-to-year yield effects were large. The permanent bed treatment with continuous crop residue retention demonstrated the highest average yield, followed by conventional tillage with residue incorporation, and then the permanent bed treatment with partial removal for fodder. Lastly, markedly reduced yields were produced under permanent beds with residue burning.

As has been observed for many long-term tillage experiments, major yield differences do not occur until several crop cycles have been planted. In the case of this experiment, no major grain yield differences were observed during the initial five years (10 harvested corps), especially at the higher N rates, which clearly highlights the need for long-term funding and institutional commitments to such trials. Had this trial been halted before the sixth year, an interpretation could have been that there were no major wheat yield differences attributable to tillage or residue management practices for this irrigated production system.

After the 1997 wheat crop, the lowest yields have been continually observed for permanent beds with residue burning. In particular for a poor yielding crop cycle like 2004, the wheat yield for permanent beds with residue burning suffered larger yield reduction as compared with the other management practices. It is clear that residue burning is not compatible with the permanent bed technology under these production conditions.

Permanent beds, with partial residue removal for fodder, have resulted in much smaller or no yield reductions in most years as compared with residue burning, especially when yield for permanent beds is compared with full residue retention. Since at least 30-40% of the loose residues and/or standing stubbles are not removed for fodder, they appear to provide adequate ground cover to benefit soil quality. In addition, the economic value of the residue removed for fodder will likely override the associated, small grain yield reductions, at least in the short term. The yield differences between permanent beds with residue retention

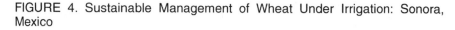

FIGURE 4. Sustainable Management of Wheat Under Irrigation: Sonora, Mexico

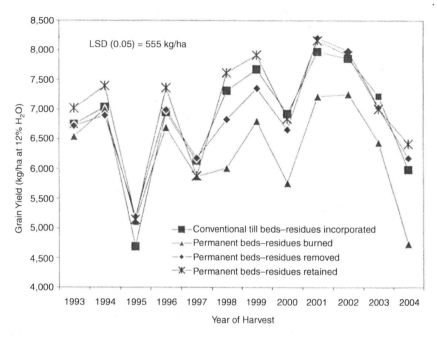

and conventional till beds with residue incorporation, were consistently small for most years. More importantly, however, the permanent beds provided a 20-30% savings in production costs.

There was only a small year by tillage/residue management interaction for wheat yield. There was no significant tillage/residue management by N management interaction although the permanent beds with crop residue retention had consistently higher yields for nearly all N management treatments, especially for the 0 N rate (data not shown). It appears that the strategy of band incorporation of both basal N and first-node stage N applications to minimize contact with the surface-retained crop residues may augment more efficient fertilizer N use.

A number of soil chemical, physical and biological parameters were regularly monitored throughout the experimental period for this long-term trial. Table 8 contains a brief summary of some of the parameters that were measured during the 2001/02 wheat-growing season. Samples that were taken at the onset of the experiment in 1992 indicated uniformity for these parameters across the trial area (data not shown).

TABLE 8. Effect of tillage and crop residue management (averaged across 4 nitrogen treatments) on soil properties for a long-term bed-planted trial initiated at Obregon, Sonora, Mexico in 1993. (Results reported here are from 0-7 cm soil samples taken during the Y2001/2002 crop cycle.)

	pH (H$_2$O)	% OM	% Total	P Olsen	Na	Fe	Zn	Wet Soil Aggregate Stability	Soil Microbial	
Tillage/ Residue Management	1.2		N	ppm	ppm	ppm	ppm	MWD#	Biomass mg C kg^{-1} soil	Biomass mg N kg^{-1} soil
Conventional Till/ Incorporate Residue	8.13	1.23	0.069	10.6	564	1.66	0.29	1.26	464	4.88
Permanent/ Beds Burn Residue	8.10	1.32	0.071	9.9	600	1.97	0.30	1.12	465	4.46
Permanent Beds/Remove Residue for Fodder	8.12	1.31	0.074	12.1	474	2.22	0.32	1.41	588	6.92
Permanent Beds/Retain Residue	8.06	1.43	0.079	10.3	448	2.03	0.44	1.96	600	9.06
Mean	8.10	1.32	0.073	10.7	513	1.97	0.34	1.44	529	6.33
LSD (P = 0.05)	0.13	0.15	0.004	5.4	53	0.31	0.12	0.33	133	1.60

Adapted from Sayre et al. 2005

The soil chemical parameters indicate that while pH was not different for the tillage/residue management treatments, Na was significantly less for permanent beds where part or all of the crop residues had been left on the soil surface. Higher levels of Na occurred for the conventional-tilled beds with residue incorporation but highest Na levels occurred for permanent beds with residue burning. This is an exceedingly important result because it indicates that soils with salinity problems could benefit from the use of permanent beds with retention of crop residues on the soil surface. Saline conditions could be ameliorated via reduction in Na accumulation in the beds as a result of reduced evaporation from the soil surface by the mulching effect of the retained residues. A reduction in Na accumulation could also be the result of improved soil physical properties that enhance Na removal by leaching action.

Organic matter levels were lowest for conventional tillage and higher for permanent beds with full crop residue retention (Table 8). It was interesting that while the soil P levels were similar across the tillage/residue management treatments, the levels of the essential micronutrients, Fe and Zn, were significantly higher for the permanent bed treatments, especially where some or all residues were retained on the soil surface.

Permanent beds with residue burning showed the poorest aggregate stability, followed by tilled beds (Table 8). Permanent beds with partial or full residue retention (especially the latter) had the best aggregate stability.

Soil microbial biomass determinations of C and N in the biomass clearly indicated the obvious superiority of permanent beds with some or all residue retention compared with permanent beds with residue burning or conventional-tilled beds with residue incorporation (Table 8). This measure of potential soil health favors the permanent beds with residue and correlates with the observations that have been made on root disease scores and soil, pathogenic nematode levels that have been consistently higher for the permanent bed treatment with residue burning (data not shown). It is evident that the inferior grain yield performance of the permanent bed treatment with crop residue burning (Figure 4) is strongly linked with the degradation associated with residue burning for most of the soil parameters that have been monitored and are considered to be associated with the sustainability of soil productivity.

Many farmers in the Yaqui Valley, Sonora, Mexico, have begun to follow wheat-maize rotation on a large commercial scale. Ing. Jorge Castro is one of those pioneering farmers who have adopted this rotation system continuously for the last 8 years. In 2004, he harvested 7 t ha^{-1} of wheat and 5 t ha^{-1} of maize with increased water-use efficiency and organic matter.

SUSTAINABLE WHEAT MANAGEMENT IN DRYLAND MEDITERRANEAN AREAS (300 MM PRECIPITATION)

ICARDA has researched sustainable wheat management systems for dryland areas since its foundation in 1977 at Tel Hadya, North Syria. In the cropping season of 2005 with precipitation of only 303 mm, the highest wheat yields of 5 ton/ha were realized on 11 ha at Tel Hadya Research Station of ICARDA by sowing early (November 16) with the spring bread wheat variety *Babaga*. The average yield of bread wheat from a larger tract of 55 ha was 4.5 tons/ha in the same season, where varieties *Cham 4* and *Babaga* were used. The possible contributing factors to this level of yield realization under water limiting conditions are listed in Table 9.

Rotation

Wheat is rotated with lentil or common vetch, with oats harvested for silage or hay on the farm. The rotation is determined by water requirement. Under rainfed conditions with a long-term average of 347 mm per season, the practice of lentil-wheat rotation is crucial. This practice can allow the storage of moisture below 25 to 30 cm in soil profile when lentil or vetch is cropped and becomes available to succeeding wheat crop.

Tillage

The heavy clay soils of the station (50 to 70 % clay, with organic matter typically ranging between 1 to 1.2 %) require, for economic reasons, no-till techniques. However, rodents and summer weed problems do not allow this option at present. Additionally, the soil is deep cracking and some form of soil mulching is required for moisture storage between growing seasons. An important aspect is the maintenance/improvement of organic matter (OM). As cereals leave more residues (3 to 6 tons/ha)

TABLE 9. Sustainable Wheat Management in Dryland Mediterranean Areas (300 mm precipitation), Tel Hadya, ICARDA, Syria.

- **Rotation:** Wheat-Legume (Lentil)
- **Soil Type:** Calcic Rhodoxeralf; 50 to 70% Clay; Organic matter 1-1.2%
- **Tillage:** 25 to 30 cm deep after Cereal; 15 cm deep after Legume
- **Residue:** 3 to 6 t/ha of Cereal and 1 to 2 t/ha of Legume
- **Sowing:** Early November
- **Cultivars:** Babaga and Cham 4
- **Seed Rate:** 100 kg/ha
- **Fertilizer:** 80 Kg N, 60 Kg P_2O_5/ha
- **Yield:** 4.5 t/ha on 50 Hectares

than legumes (1 to 2 tons/ha) and also because the loss of moisture is lower after cereals than after legumes, deep tillage with a moldboard plough (25 to 30 cm) is carried out after cereals. Straw incorporation (about 80%) across a decade has raised the organic matter content from often below 1% to 1.2%.

After legume harvest, a duck-foot cultivator carries out tillage to a maximum depth of 15 cm, preferably only 10 to 12 cm. The purpose is to control summer weeds, like Bermuda grass (*Cynodon dactylon* (L.) Pers.), Johnsongrass (*Sorghum halepense* (L.) Pers.), liquorice (*Glycyrrhiza glabra* L.), and also late thistles and other weeds. Two passages are required in most years for effective weed control. Both tillage systems bury the residues, seeds as well as straw, to make it difficult for rodents (mostly voles) to collect feed and nesting material. Another effect is to secure the residues from blowing away in strong, gusty summer winds. Most importantly, however, the shallow cultivation, following the legumes, seals the deep soil cracks, providing a soil-mulch for residual water storage.

Sowing

The most important factors for maximum production are:

* Timely sowing
* Shallow sowing
* Narrow row spacing

Under the conditions of northern Syria, timely sowing of wheat means the first half of November. The later the sowing, the higher the risk of drought and heat stress at the end of the season. Stress due to late frost is relatively rare.

Shallow sowing is required, contrary to farmers' usual practice. Farmers have problems achieving an accurate planting depth, because of a lack of any seedbed preparation equipment, other than a duck-foot cultivator. As a consequence, the seeders are unable to place seeds in a uniform, shallow depth of 3 to 5 cm.

Row spacing should be kept at the minimum possible. At a row spacing of > 20 cm, plants are unable to form a closed leaf canopy early. Where residues, seedbed and equipment permit, 10 to 15 cm row spacing is achievable.

Cultivar

At the ICARDA station, two varieties are widely grown at present:

- *Cham 4*, spring bread wheat, is about 15 years old. Both are spring bread wheats and are resistant to the important prevailing pests and diseases. Besides seed dressing, no other fungicide/insecticide is applied.
- *Babaga*, available since 2002, is also spring bread wheat and is characterized by wider leaves that shade the ground early and to a higher degree than *Cham 4*. This helps reduce surface evaporation during the later phase of the season, when the risk of drought is substantial.

Seed Rate

The standard seed rate has been 100 kg/ha for early sowing and up to 120 kg/ha for later planted wheat (December). Farmers sow 200 to 250 kg seed/ha at present due to lack of proper seedbeds. With Syrian wheat prices of $180-200/ton for bread wheat, this results in an additional cost of $20 to 30/ha for the higher seed rate. This justifies investment into better seedbed preparation and planting techniques.

Fertilization

Fertilizer requirement is limited to N and P. The N-fertilization is carried out with urea, 40 kg broadcast before sowing and 40 kg at tillering, but under drier conditions the top dressing may be reduced to 30 kg/ha. This low N-application level is sufficient for the typical rotation with 50% legumes with straw incorporation.

Another major contributing factor to these high dryland yields was the absence of heat stress prior to grain filling in the 2004/2005 season. There were only three days with temperatures > 30°C before May 10, 2005.

SUSTAINABILITY THROUGH DROUGHT-TOLERANT GENOTYPES

Water-use efficiency in agriculture is one of the most important issues for consideration in the future, especially in dry areas in Central and West Asia and North Africa (CWANA). Vast areas of the Indian

Subcontinent and China are irrigated; nonetheless, the amount of water needed for agriculture is getting to be in short supply due primarily to industrialization and population pressure. In many parts of the world where dryland agriculture is practical, e.g., South America, Australia and Southern Africa, the rainfall pattern has become erratic due to global warming and deforestation. In many places, the soils are degraded and have lost the water holding capacity.

Drought-tolerant varieties of wheat and other crops can be a component of the solution to the water-deficit problem. There is continuous debate on the physiological mechanism of drought tolerance and methodology in breeding. The physiological factors of plants, soil types, and rainfall amount and pattern, together influence crop performance in a given environment. Without any doubt, the genetics influences drought tolerance magnitude; genotype-by-environment interaction also has a strong influence on drought tolerance magnitude. Experimental results have conclusively demonstrated the superiority of some genotypes over others.

During the past 50 years, a lot of efforts and resources have been invested in classification, investigation of genetic mechanisms, physiological characterization and breeding methodology for drought tolerance. The genetic gains for yield improvement in semiarid environments have been substantial under some circumstances (Pingali and Rajaram, 1999) and poor in others (Perry and D'Antuono, 1989).

Recently Rajaram (1999) suggested that an alternative methodology of breeding for adaptation to drought conditions under reduced irrigation and high yield performance under optimum amount of water and fertilization, denoted as yield potential, be combined into a single genotype. Rajaram (senior author) and colleagues have seen such performance in genotypes such as variety Attila in international environments of Asia and North Africa, including India and Ethiopia. Such genotypes have given high performance in high yielding environments while maintaining high yields under drought stress conditions. This phenomenon is graphically presented in Figure 5 relating to data produced at Tel Hadya, Syria (ICARDA) under rainfed condition (average 300 mm of precipitation) (data provided by Dr. Osman Abdalla,[†] unpublished). The varieties 506, 510, 508, 505, 511, and 512 are outstanding under moisture stress (280 mm), yielding above 1.5 t ha^{-1}. Some poorer genotypes yielded only 500 kg ha^{-1} in this experiment. The listed genotypes performed very well under improved moisture provided as supplemen-

---

[†]Abdalla, O. Performance of spring bread wheat cultivars under moisture stress extremes (unpublished data used with permission).

FIGURE 5. Drought Tolerance + Yield Potential Genotypes: ICARDA, Syria, 2003/2004

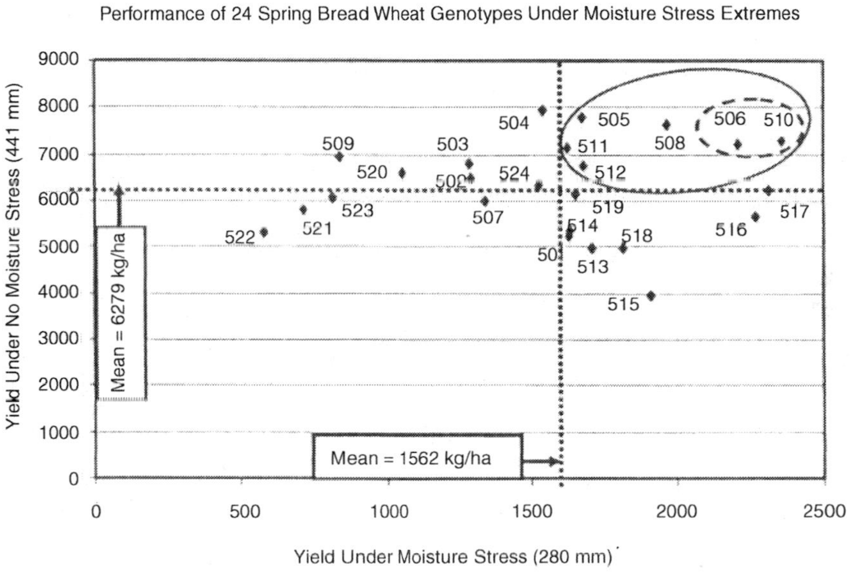

Performance of 24 Spring Bread Wheat Genotypes Under Moisture Stress Extremes

Source: Osman Abdalla (unpublished), ICARDA

tary irrigations (441 mm) and produced yields between 6 t ha^{-1} and 7.8 t ha^{-1}. This is not the case of yield potential being translated into high yield under drought environment. It is very clear that certain genotypes, e.g., 504, 509, and 503, gave the highest yields of above 7 t ha^{-1}, however, their performance under 280 mm of rainfall was less than 1500 kg ha^{-1}. Genotype 509 gave less than 1 t ha^{-1} under 280 mm rainfall. These results and many others published by the senior author prove the validity of this concept and we recommend this modified breeding methodology to achieve the desired result.

REFERENCES

Anthony, G. and J.P. Brennan. 1987. Progress in yield potential and bread making characteristics in wheat in New South Wales. 1925-26 to 1984-85. Agricultural Economics Bulletin, Division of Marketing and Economic Services, New South Wales, Australia: Department of Agriculture.

Blevins, R.L., M.S. Smith and G.W. Thomas. 1984. Changes in soil properties under no-tillage. In: Phillips, R.E., and S.H. Phillips (Eds.). No-Tillage Agriculture–Principles and Practices. Pp. 190-230. Van Nostrand Rheinhold Company, New York, USA.

Byerlee, D. and P.F. Moya. 1993. Impacts of international wheat breeding research in the developing world. 1966-1990. Mexico, D.F.: CIMMYT.

Cox, T.S., J.P. Shroyer, B.H., Liu, R.G. Sears and T.J. Martin. 1988. Genetic improvement in agronomic traits of hard red winter wheat cultivars from 1919 to 1987. Crop Science. 28: 756-760.

Kronstad, W.E. 1998. Agricultural development and wheat breeding in the 20th century. Pp. 1-10. In: H.J. Braun, F. Altay, W.E. Kronstad, S.P.S. Beniwal, and A. McNab (eds). Wheat: Prospects for Global Improvement. Proc. of the 5th Int. Wheat Conf., Ankara, Turkey. Developments in Plant Breeding, v. 6. Kluwer Academic Publishers. Dordrecht.

Kulshrestha, V.P. and H.K. Jain. 1982. Eighty years of wheat breeding in India: Past selection pressure and future prospects. Zeitschrift fur Pflanzenzuchtung. 89: 19-30.

Lantican, M.A., H.J. Dubin and M.L. Morris. 2005. Impact of International Wheat Breeding Research in the Developing World, 1088-2002. Mexico, CIMMYT.

Mashiringwani, N.A. 1987. Trends in production and consumption of wheat and role of variety improvement in Zimbabwe. Department of Research and Specialist Services.

Mitchel, D.O., M.D. Onco and R.D. Duncan. 1989. The World Food Outlook. Cambridge University Press.

Nagarajan, S. 2002. The significance of wheat production for India, in particular under limited moisture conditions. http://www.biotech.boil.ethz.ch/India/forms.Nagarajan.pdf.

Perry, M.W. and M.F. D'Antuono. 1989. Yield improvement and associated characteristics of some Australian spring wheat cultivars introduced between 1860 and 1982. Australian Journal of Agricultural Research. 40: 457-472.

Pingali, P.L. and S. Rajaram. 1999. Global wheat research in a changing world: options for sustaining growth in wheat productivity. In: P.L. Pingali (ed.). CIMMYT 1998-99 World Wheat Facts and Trends. Global Wheat Research in a Changing World: Changes and Challenges, Mexico, CIMMYT.

Rajaram, S., M. Van Ginkel and R.A. Fischer. 1994. CIMMYT's Wheat Breeding Mega environments (ME). In: Proceeding of the 8th International Wheat Genetic Symposium, Beijing, China. Pp. 1101-1106.

Rajaram, S. 1999. Half a Century of international wheat breeding. In: Behl, R.K., M.S. Purina and B.P.S. Lather (Eds.). Crop Improvement for Food Security. SSARM, Hissar. Pp. 158-183.

Rosegrant, M.W., A. Agcaolli-Sombilla and N. Perez. 1995. Global Food Projections to 2020. IFPRI, Washington, DC.

Ruttan, V.W. 1993. Research to meet crop production needs: Into the 20th Century. In: D.R. Buxton et al. (eds.) International Crop Science Congress. CCSA, Madison, Wisconsin, USA. Pp. 3-10.

Sayre, K.D., S. Rajaram and R.A. Fischer. 1999. Yield potential progress in short bread wheats in Northwest Mexico. Crop Science. 37(1): 36-42.

Sayre, K.D., A. Limon-Ortega, B. Govaerts, A. Martinez and M. Ruiz Cano. 2005. Effects following twelve years of irrigated permanent raised bed planting systems in North West Mexico. Paper presented at the ISTRO International Conference on Land Use at Brno, Czech Republic. July 1-3, 2005.

Silvey, V. 1978. The contribution of new varieties to increasing cereal (wheat and barley) yields in England and Wales. Journal of the National Institute of Agriculture Botany. 14(3): 367-384.

Tomassini, R.G.A. 2002. Impacts of Mexican germplasm on Brazilian wheat cropping: An ex-post economic analysis. CIMMYT Economica Working Paper No. 02-01. Mexico, D.F.: CIMMYT.

Van Lill, D. and J.L. Purchase. 1995. Directions in breeding for winter wheat yield and quality in South Africa from 1930 to 1990. Euphytica. 82: 79-87.

Waddington, S.R., J.K. Ranson, M. Osmanzai and D.A. Saunders. 1986. Improvement in the yield potential of bread wheat adapted to Northwest Mexico. Crop Science. 26: 698-703.

doi:10.1300/J411v19n01_05

Sustainability
of the Rice-Wheat Cropping System:
Issues, Constraints, and Remedial Options

J. K. Ladha

H. Pathak

R. K. Gupta

SUMMARY. The rice-wheat cropping system of the Indo-Gangetic Plains (IGP) has contributed tremendously to food security of the region. However, of late there has been a significant slowdown in yield growth rate of this system and the sustainability of this important cropping system is at stake. A decline in soil productivity, particularly of organic C and N, a deterioration in soil physical characteristics, a delay in sowing of wheat, and decreasing water availability are often suggested as the causes of this slowdown in productivity. Therefore, a paradigm shift is

J. K. Ladha is a Representative, International Rice Research Institute, India Office, CG Block, NASC Complex, DPS Marg, Pusa Campus, New Delhi-110012, India.

H. Pathak is Senior Associate Scientist, International Rice Research Institute, India Office, CG Block, NASC Complex, DPS Marg, Pusa Campus, New Delhi-110012, India.

R. K. Gupta is affiliated with the Rice-Wheat Consortium for IGP, CIMMYT-RWC, CG Block, NASC Complex, DPS Marg, Pusa Campus, New Delhi-110012, India.

Address correspondence to: J. K. Ladha at the above address (E-mail: J.K.Ladha@cgiar.org).

[Haworth co-indexing entry note]: "Sustainability of the Rice-Wheat Cropping System: Issues, Constraints, and Remedial Options." Ladha, J. K., H. Pathak, and R. K. Gupta. Co-published simultaneously in *Journal of Crop Improvement* (Haworth Food & Agricultural Products Press, an imprint of The Haworth Press, Inc.) Vol. 19, No. 1/2 (#37/38), 2007, pp. 125-136; and: *Agricultural and Environmental Sustainability: Considerations for the Future* (ed: Manjit S. Kang) Haworth Food & Agricultural Products Press, an imprint of The Haworth Press, Inc., 2007, pp. 125-136. Single or multiple copies of this article are available for a fee from The Haworth Document Delivery Service [1-800-HAWORTH, 9:00 a.m. - 5:00 p.m. (EST). E-mail address: docdelivery@haworthpress.com].

Available online at http://jcrip.haworthpress.com
doi:10.1300/J411v19n01_06

required for enhancing the system's productivity and sustainability. Resource-conserving technologies involving zero- or minimum tillage with direct seeding, improved water-use efficiency, innovations in residue management to avoid straw burning, and crop diversification should assist in achieving sustainable productivity and allow farmers to minimize inputs, maximize yields, conserve the natural resource base, reduce risk due to both environmental and economic factors, and increase profitability. doi:10.1300/J411v19n01_06 *[Article copies available for a fee from The Haworth Document Delivery Service: 1-800-HAWORTH. E-mail address: <docdelivery@haworthpress.com> Website: <http://www.HaworthPress.com> © 2007 by The Haworth Press, Inc. All rights reserved.]*

KEYWORDS. Indo-Gangetic Plains, N-use efficiency, resource-conserving technologies, soil fertility, water-use efficiency, yield stagnation

INTRODUCTION

Four decades after the Green Revolution (GR), which made South Asia food-secure, the sustainability of agriculture is posing a serious challenge to the region's food security all over again. Today, scientists are trying to understand where the GR went wrong. The answer probably lies in the fact that, at the time of the GR, increase in yield was the only criterion, with little or no emphasis on sustainability. The way to increase yield was to increase inputs, such as fertilizer and pesticides, ignoring their long-term effects on the natural resource base and the environment. In the GR, rice (*Oryza sativa* L.) and wheat (*Triticum aestivum* L.) became the most important crops with thirsty and hungry high-yielding cultivars. Though the immediate food demand was met, an increased demand for water led to over-extraction of groundwater, indiscriminate intensive cropping robbed the soil of its nutrients, and the non-judicious use of chemicals poisoned the groundwater and contaminated crops.

Rice and wheat account for more than 80% of the total cereal production in the Indo-Gangetic Plains (IGP). The intensively cultivated irrigated rice-wheat system is the basis of employment, income, and livelihood for hundreds of millions of rural and urban poor of South Asia. In the last few decades, high growth rates for food grain production (3.0% for wheat and 2.3% for rice) in the IGP have kept pace with population growth. But recent evidence shows that productivity of the rice-wheat system has been plateauing because of a fatigued natural resource base. Thus, the region's food security is once again threatened. In addition, in many of the eastern areas of the IGP, which are located in unfavorable rice ecologies, gains from GR technologies seem not to have affected the lives of millions. Problems of the rice-wheat cropping system in the different transects of the IGP are summarized in Table 1.

ISSUES AND CONSTRAINTS

Demand for rice and wheat in South Asia is expected to grow at 2.02% and 2.49% per year, respectively, during the next couple of decades (Rosegrant et al., 2002), requiring continuing efforts to increase production and productivity. Hobbs and Gupta (2003) reported a decline in the per capita land area of the rice-wheat system from 1200 m² in 1961 to 700 m² at present. This decline is likely to continue as the population and competition from other crops, urbanization, and industrialization continue to increase.

There has lately been a significant slowdown in the growth rate of production as well as yield in the IGP (Sinha et al., 1998; Ladha et al., 2003; Pathak et al., 2003). Padre and Ladha (2006) observed a declining trend in rice yield in rice-wheat long-term experiments in South Asia, including China, with recommended rates of nutrients. Rice yields declined throughout Asia in the wet season, but there has been no significant change in wheat yields. Pathak et al. (2003) estimated the potential yield trend of rice and wheat at different sites inside and outside the IGP in India. The potential yield trend in rice ranged from -0.12 Mg ha^{-1} yr^{-1} at Delhi to 0.05 Mg ha^{-1} yr^{-1} at Kanpur. Negative yield trends were observed in 6 of the 9 data sets, 4 of which were significantly different from 0 ($P < 0.05$). These declining trends were observed across transects of the IGP, indicating that yield decline is not localized. On the other hand, positive but statistically nonsignificant trends were observed at 3 sites. In wheat, the annual yield trend ranged from -0.07 Mg ha^{-1} yr^{-1} at Delhi to 0.04 Mg ha^{-1} yr^{-1} at Faizabad and Pantnagar. Of the 9 sites, 6 showed a negative trend whereas 3 showed a positive trend but none was significantly different from 0. Thus, wheat yield appeared to be more stable than rice yield. Ladha et al. (2003) examined the possible causes of the yield decline and observed that degradation of the soil and water resource base and inefficient nutrient management could be responsible for this trend (Table 2).

In most of the rice-wheat areas, the soil organic matter content has declined across time, with a corresponding yield decline of rice and

TABLE 1. Problems of the Rice-Wheat Cropping System in the Different Transects of the Indo-Gangetic Plains (IGP)

Factor	IGP1 and 2[a]	IGP3	IGP4	IGP5
Water	Shortage of good-quality irrigation water and low water-use efficiency	Shortage of irrigation water	Shortage of irrigation water and occasional flood and drought	Waterlogging, flood, excessive soil moisture in wheat
Soil	Low soil organic matter, rising salinity	Low soil organic matter, rising salinity, alkalinity	Nutrient mining	Nutrient mining, rising salinity
Fertilizer	Inefficient N use, imbalanced fertilizer use	Inefficient N use, imbalanced fertilizer use	Low fertilizer use	Imbalanced fertilizer use
Crop management	Delay in sowing of wheat, difficulty in residue management	Poor land leveling, low plant density in rice, late sowing of wheat	Non-adoption of high-yielding varieties, poor land leveling, low mechanization	Low mechanization, poor land leveling
Climate	-	Rain during maturity of rice	Rise in temperature during grain filling of wheat, shorter wheat season, rain and storm during maturity of rice	Rain and storm during maturity of rice and wheat, shorter wheat season
Labor	Unavailable when needed	Unavailable when needed	-	Unavailable when needed

[a]IGP1 and IGP2, areas in Pakistan and parts of Punjab and Haryana in India; IGP3, most of Uttar Pradesh and parts of Bihar, India, and parts of Nepal; IGP4, parts of Bihar, India, and parts of Nepal, IGP5, parts of Bihar and West Bengal, India, and parts of Bangladesh.
Adapted from Ladha et al. (2003)

TABLE 2. Causes of Yield Decline in the Rice-Wheat System in the Indo-Gangetic Plains

- Decline in soil organic C, N, P, K, and Zn
- Long-term changes in soil physical characteristics
- Delay in sowing of wheat
- Decreasing water availability
- Salinity and water logging
- Increased pest incidence and evolution of new, more virulent pests
- Groundwater depletion
- Varietal substitution
- Changes in land quality (sodicity)
- Increased weed infestation and herbicide resistance
- Decrease in solar radiation and increase in minimum temperature

wheat at recommended rates of NPK application (Ladha et al., 2003). In the major rice-wheat regions of Northwestern India, the soil organic C (SOC) has dropped from 0.5% in the 1960s to 0.2% in the 1990s (Sinha et al., 1998). A similar decline in SOC was evidenced in the Punjab in Pakistan (Ali and Byerlee, 2000). During 1971-74, SOC was 1.02%, which decreased to 0.72% during 1975-84 and further declined to 0.59% in the post-GR period of 1985-94. Non-recycling of organic matter is the principal cause of this decline. After the grain harvest, the stubble is sometimes removed to be used as forage, fuel, or building material. Stubble is also burned to facilitate land preparation for a subsequent rice crop. In areas where harvesting is done with a combine, e.g., Northwestern India, rice straw is generally burned before planting the next crop (Samra et al., 2003). Only roots and sometimes stubble are recycled.

Other changes in soil, such as changes in soil physical properties, formation of a hardpan in subsoil, accumulation of toxic substances, and changes in soil microbial biomass, have not been periodically studied, and therefore it is difficult to assess their role in the yield decline. However, it can be concluded that soil fertility degradation is a problem in the rice-wheat system, but the specific form it takes varies greatly from one place to another.

In the IGP, water availability for irrigation has decreased because of more area having been planted to rice, which reduced water availability per unit area (Sinha et al., 1998). A reduced water supply through canal irrigation is also linked to less water-reserving capacity of the reservoirs due to increasing siltation and increasing competition for water for domestic and industrial use (Hobbs and Gupta, 2003). Soils of the upper transect of the IGP are light to medium textured and a large amount of water (> 250 cm) is required for rice production (Narang and Singh, 1988). Since rainfall is not sufficient to meet the water requirement, frequent irrigations are needed. Canal water accounts for 35 to 40% of the total irrigation requirement and the remaining requirement is met from groundwater. Some progressive farmers in this region have been growing summer rice (April-June), along with rice and wheat in the main season (Aggarwal et al., 2000). The additional groundwater required for summer rice can further imbalance the groundwater budget. Pumping from deeper layers increases total operational costs and thus decreases the profitability of growing rice. This has already resulted in a decline in water aquifers and water quality in many regions (Sinha et al., 1998; Singh, 2000). Reportedly, during the last two decades, the groundwater table has fallen at a rate of about 23 cm yr^{-1} in the Central Punjab

(World Bank, 2003). There are also signs that the rate of depletion may have become more pronounced in recent years because of a larger area under rice cultivation and the practice of transplanting rice early.

The other side of the water problem is waterlogging in some areas in the trans-Gangetic plains in India. An impressive canal network in this region has made possible the intensive rice-wheat cropping system, but with water percolation from canals, distributaries, and water-courses, the water table has risen in some areas. Use of excess amounts of irrigation water is another cause of the water-table rise.

The drastically different seedbed requirements for rice and wheat create problems in tillage, the timeliness of wheat sowing, the maintenance of soil structure, and the management of irrigation water, weeds, other pests, fertilizers, and crop residues. The short turnaround time between rice and wheat and the insistence of farmers practicing excessive preparatory tillage delay wheat planting, which results in yield losses of 35 kg day^{-1} ha^{-1} in the northwest and up to 60 kg day^{-1} ha^{-1} in the eastern IGP (Hobbs and Gupta, 2003). Also, many areas in the eastern IGP remain as "rice fallows" because fields remain wet for a long time, forcing farmers to wait before any preparatory tillage is possible for planting other crops. Wheat sowing is also delayed because of the planting of medium-duration (140 days) basmati rice. In the eastern IGP, farmers begin preparatory tillage for the rice nursery and transplant seedlings in the main fields after the onset of the monsoons, which results in low crop yields. This practice also wastes 400 to 600 mm of rainwater.

The disposal of rice straw is a major concern in places where the rice-wheat cropping system is extensively followed. Rice straw is not used as animal feed because of its low digestibility and low protein, high lignin, and high silica contents. It is also not recycled in the soil because of the limited time (20 to 25 days) left before sowing of the succeeding wheat crop. Within this short period rice, straw cannot be completely decomposed in the soil. Moreover, due to the addition of a large amount of organic C of poor quality through rice straw, a net immobilization of N occurs in the soil and the wheat crop suffers from N deficiency, resulting in lower yield. Farmers in northwest India therefore dispose of a large part of the rice straw by burning it *in situ*. In a recent survey, it was noted that 60% and 82% of the rice straw produced in the northwestern states of Haryana and Punjab, respectively, was burned in the field (Punjab Agricultural University, unpublished).

Diagnostic surveys on nutrient management practices prevailing in rice-wheat dominated areas revealed that nearly one-third of rice-wheat farmers apply as much as 180 kg fertilizer N ha^{-1} to each rice and wheat crop vis-à-vis the local recommendation of 120 kg N ha^{-1}. As a result, N-recovery efficiency, which is already not more than 50% in the irrigated rice-wheat system (Ladha et al., 2005), has declined further because of irrational N applications (Bijay-Singh et al., 2002). Nitrogen applied in excess of the crop's demand is lost through various pathways, and this has a negative effect on the environment and reduces economic benefits to farmers. Management of nutrients should be given adequate attention to increase yields and sustain productivity.

ENHANCING THE SUSTAINABILITY OF THE RICE-WHEAT SYSTEM

Resource-conserving technologies, also termed as "conservation agriculture," involving zero- or minimum tillage with direct seeding, and bed planting with residue mulch, are being advocated as alternatives to the conventional rice-wheat system for improving sustainability. The performance of rice on beds in the IGP has been variable. With similar irrigation scheduling, yields on permanent beds are generally lower than in puddled transplanted rice, with problems of iron deficiency, weeds, accurate sowing depth, and sometimes nematodes, particularly in direct-seeded rice on beds. All these problems lead to poor adoption of this technology by farmers. Strategies for overcoming the problems are urgently needed, including breeding and selection for rice grown under aerobic soil conditions and for wide-row spacing between adjacent beds.

No-tillage or minimum tillage on flat land in the rice-wheat system is becoming attractive to farmers in Asia because of higher input-use efficiency and lower cost of cultivation. For example, when N was deep-placed in rice and wheat under no-tillage, system yields were maintained and the recovery efficiency of N (RE_N) increased (Ladha et al., 2003). Wheat yield increased with timely planting of the crop under reduced tillage in South Asia (Hobbs and Gupta, 2003). Because the amount of fertilizer N applied does not generally increase in those systems, RE_N also increases. In southern Brazil, after 7 years of zero-tillage, organic matter in surface soil (0-10 cm) increased significantly and the N rate in maize (*Zea mays* L.) for a yield goal of 7 Mg ha^{-1} decreased from 150 to 75 kg N ha^{-1} starting with the 5th year after the introduction of zero-tillage, which suggested a strong improvement in

RE_N (Boddey et al., 1997). Although crop yields are more or less the same under conservation/zero- and conventional tillage, conservation/zero-tillage often proves superior in erosion control, resource use, and operational cost. Reduced loss and increased availability of applied N under zero-tillage enhanced yield and N-use efficiency in winter wheat (Rao and Dao, 1996).

The productivity of water for rice is considered as very low in general and other crops are considered to be more productive than rice in water use. Although there is considerable scope for enhancing the water productivity of rice-based systems, more systematic evaluations are needed to understand the water balance dynamics and productivity of water at various scales of the irrigation system. Specific to the rice-based system facing water scarcity problems, the improvement in productivity of water at the field/farm scale will mainly result through (1) increasing the yield per unit of water that is transpired; (2) reducing unproductive seepage, percolation, and evaporation outflows; and (3) making more effective use of rainfall (Tuang and Bouman, 2003).

Nitrogen-use efficiency (NUE) can be improved by adopting fertilizer, soil, water, and crop management practices that will maximize crop N uptake, minimize N losses, and optimize indigenous soil N supply. Management decisions that increase fertilizer-N use by crops can focus on two approaches: (1) increase fertilizer-N use during the growing season when the fertilizer is applied and (2) decrease fertilizer-N losses, thereby increasing the potential recovery of residual fertilizer-N by the subsequent crops. Removing plant-growth-limiting factors would increase crop demand for N, leading to a greater use of available N and consequently higher NUE (Ladha et al., 2005).

Residue management may be an important factor in improving soil physical, chemical, and biological properties and maintaining yields in the rice-wheat system. With widespread mechanization and the use of combine harvesters in northwest India, the majority of rice residues and significant amounts of wheat straw are burned in the field, creating serious air pollution in addition to the loss of nutrients and soil organic matter (Yadvinder-Singh et al., 2005). Adoption of stubble retention requires development of machinery with the capability of sowing into rice residues. Considerable progress has been made in developing technology for direct drilling into stubbles on flatland, including double and triple disc assemblies and the star-wheel punch planter (RWC, 2002), and the "Happy Seeder" (Sidhu et al., 2004). The Happy Seeder combines stubble-mulching and seed-drilling functions into one machine. Versions are available that can be used on flatland or in bed layouts by simply changing the seed drill behind the mulcher. The impact of mulching on components of the water balance for the rice-wheat systems needs to be evaluated.

Diversification, i.e., growing a range of crops suited to different sowing and harvesting times, assists in achieving sustainable productivity by allowing farmers to employ biological cycles to minimize inputs, maximize yields, conserve the resource base, and reduce risk from both environmental and economic factors. This also enables farmers to manage larger areas while attending to each crop at optimal times. To reverse or arrest this downward trend of sustainable productivity of the IGP, a substantial change in the current cropping system is required, including reducing tillage and improving organic matter status (Connolly et al., 2001). The significance of crop diversification was realized in Australia for sustaining the crop productivity and environment (Connolly et al., 2001) because agricultural diversification is considered to have important implications for income and employment in rural areas and stabilizing effects on the environment (Bhatia and Tiwari, 1999). Further, the nature and extent of crop diversification have not witnessed any conflict with self-sufficiency (Pandey and Sharma, 1996). The farmers of the rice-wheat belt have taken the initiative to diversify their farming systems by including short-duration crops, for example, potato, soybean, urd, mungbean, cowpea, pea, mustard, maize, etc., in different combinations.

In Asian countries, an estimated 25-50% of the total grain value is lost through poor timing of harvesting and threshing, inadequate grain moisture control at various stages of processing, and inefficient grain handling and milling. Farmers may be aware of the losses in quantity of grain (10-15%) but not the losses in grain quality (10-35%), which determine the market price for their produce. A majority of farmers allow the rice crop to over-ripen before harvest and wait for 2 to 5 days between harvest and threshing. These practices increase grain breakage in milling and reduce whole-grain recovery and the price (value) for milled rice. In other situations, farmers harvest the crop and keep it in the main field to dry before processing. Education of farmers in harvest and post-harvest handling and the use of suitable tools, such as the moisture meter, milling degree chart, milling equipment, etc., will minimize losses in quantity and quality at all levels. An improvement in harvest and post-harvest processing will enhance the market value of rice and thus improve farmers' income and livelihood, and enhance millers' profits.

CONCLUSIONS

Despite tremendous gains in rice and wheat production during the last 40 years, the continued growth in population and pressure on the natural resource base place an extreme stress on the rice-wheat system. Additional gains in productivity, profit, and product quality are becoming increasingly difficult to achieve by using the single-technology-centric approach. Therefore, a systems approach is necessary to increase the productivity of both rice and wheat crops grown in sequence in the IGP. Currently, most rice-wheat farmers have limited access to new knowledge, technologies, and options that could help them achieve sustainable gains in production and improvements in livelihoods. For farmers, this "knowledge gap" results in low profitability, poor standard of living, and inadvertent environmental damage. There is a need to build on the scientific and technological gains made recently by various national and international partners in integrating all available rice and wheat production technologies, evaluating them in farmers' fields, and promoting the successful ones to farmers.

REFERENCES

Aggarwal, P.K., S.K. Bandyopadhyay, H. Pathak, N. Kalra, S. Chander and S. Sujith Kumar. (2000). Analyses of yield trends of the rice-wheat system in north-western India. Outlook Agric. 29:259-268.

Ali, M. and D. Byerlee. (2000). Productivity growth and resource degradation in Pakistan's Punjab: A decomposition analysis. Policy Research Working Paper 2480, World Bank, Washington, DC.

Bhatia J. and S.K. Tewari. (1999). Diversification growth and stability of agricultural economy in Uttar Pradesh. Agric. Situation India 45:397-403.

Bijay-Singh, P.R. Gajri, J. Timsina, Yadvinder-Singh and S.S. Dhillon. (2002). Some issues on water and nitrogen dynamics in rice-wheat sequences on flats and beds in the Indo-Gangetic plains. In: Humphreys, E. and Timsina, J. (eds.) Modelling irrigated cropping systems, with special attention to rice-wheat sequences and raised bed planting, CSIRO Land and Water Tech. Report 25/02. Proc. Workshop CSIRO, Griffith, 25-28 Feb. 2002, pp. 1-15.

Boddey, R.M., J.C. Sa, B.J. Alves and S. Urquiaga. (1997). The contribution of biological nitrogen fixation for sustainable agriculture systems in the tropics. Soil Biol. Biochem. 29:787-799.

Connolly R.D., D.M. Freebairn, M.J. Bell and G. Thomas. (2001). Effects of rundown in soil hydraulic condition on crop productivity in south-east Queensland–A simulation study. Austr. J. Soil Res. 39:1111-1129.

Hobbs, P. R. and R.K. Gupta. (2003). Resource-conserving technologies for wheat in the rice-wheat system. In: Improving the productivity and sustainability of rice-wheat system: issues and impacts. J.K. Ladha, J.E. Hill, J.M. Duxbury, R.K. Gupta, and R.J. Buresh (Eds.), ASA Special Publication 65, Madison, WI, USA, pp. 149-171.

Ladha, J.K., H. Pathak, T.J. Krupnik, J. Six and C. van Kessel. (2005). Efficiency of fertilizer nitrogen in cereal production: Retrospect and prospects. Adv. Agron. 87:85-156.

Ladha, J.K., H. Pathak, A.T. Padre, D. Dawe and R.K. Gupta. (2003). Productivity trends in intensive rice-wheat cropping systems in Asia. In: Improving the productivity and sustainability of rice-wheat systems: Issues and impacts. J.K. Ladha, J.E. Hill, J.M. Duxbury, R.K. Gupta and R.J. Buresh (Eds.), ASA Special Publication 65, ASA-CSSA-SSSA, Madison, WI, USA, pp. 45-76.

Narang, R.S. and N. Singh. (1988). Water requirement studies in rice. Inter. Rice Res. News 13:43.

Padre, A.T. and J.K. Ladha. (2006). Integrating rice and wheat productivity trends using the SAS mixed-procedure and meta-analysis. Field Crops Res. 95:75-88.

Pandey V.K. and K.C. Sharma. (1996). Crop diversification and self-sufficiency in food grains. Ind. J. Agric. Econ. 51:644-651.

Pathak, H., J.K. Ladha, P.K. Aggarwal, S. Peng, S. Das, Yadvinder Singh, Bijay-Singh, S.K. Kamra, B. Mishra, A.S.R.A.S. Sastri, H.P. Aggarwal, D.K. Das and R.K. Gupta. (2003). Climatic potential and on-farm yield trends of rice and wheat in the Indo-Gangetic plains. Field Crops Res. 80(3):223-234.

Rao, S.C. and T.H. Dao. (1996). Nitrogen placement and tillage effects on dry matter and nitrogen accumulation and redistribution in winter wheat. Agron. J. 88(3): 365 371.

Rosegrant, M.W., X. Cai and S.A. Cline. (2002). World water and food to 2025: Dealing with scarcity. International Food Policy Research Institute (IFPRI), Washington, DC.

RWC. (2002). Design improvements in existing zero-till machines for residue conditions. Rice-Wheat Consortium Traveling Seminar Report Series 3 (RWC-CIMMYT, New Delhi, India).

Samra, J.S., Bijay-Singh and K. Kumar. (2003). Managing Crop Residues in the Rice-Wheat System of the Indo-Gangetic Plain. In: Improving the productivity and sustainability of rice-wheat systems: Issues and impacts. J.K. Ladha, J.E. Hill, J.M. Duxbury, R.K. Gupta and R.J. Buresh (Eds.), ASA Special Publication 65, ASA-CSSA-SSSA, Madison, WI, USA, pp. 145-176.

Sidhu, A.S., S.S. Kukal, D. Singh and V.K. Singh. (2004). Reducing water and nutrient losses in rice grown in different soils. Final Report, National Agricultural Technology Project, Department of Soils, Punjab Agricultural University, Ludhiana, India.

Singh, R.B. 2000. Environmental consequences of agricultural development: A case study from the Green Revolution state of Haryana. Agric. Ecosyst. Environ. 82:97-103.

Sinha, S.K., G.B. Singh and M. Rai. (1998). Decline in crop productivity in Haryana and Punjab: Myth or reality? Indian Council of Agricultural Research, New Delhi, 89 pp.

Tuang, T.P. and B.A.M. Bouman. (2003). Rice production in water scarce environments. In: Water productivity in agriculture: Limits and opportunities for improvement. Kijne, J.W., Barker, R., and Molden, D. (Eds.), CABI Publishing in association with the International Water Management Institute.

World Bank. (2003). India Revitalizing Punjab's Agriculture, Rural Development Unit, South Asia Region, The World Bank.

Yadvinder-Singh (2005). Progress report (2003). IAEA Coordinated Research Project on Integrated Soil, Water and Nutrient Management for Rice-Wheat Cropping Systems. International Atomic Energy Agency, Vienna.

doi:10.1300/J411v19n01_06

Tailoring Conservation Agriculture to the Needs of Small Farmers in Developing Countries: An Analysis of Issues

Patrick C. Wall

SUMMARY. Conservation agriculture (CA) is characterized by surface crop residue retention and minimal soil movement. It is a complex technology that involves not only a change in many of the farmer's cultural practices, but also a change in mind-set to overcome the use of the plow. CA is knowledge-intensive, and success with the system may depend more on what the farmer does than the level of inputs applied. However, smallholder farmers are generally characterized by weak links to information systems outside those of the community, while close community linkages tend to reinforce traditional activities. They commonly manage complex crop-livestock systems, where crop residues play an important role in animal nutrition. This, coupled with communal grazing rights, makes the retention of sufficient surface residues a difficult practice for small farmers, especially in rain-fed systems where residue production levels are low. Smallholders often have weak links to input and output markets, as well as limited access to capital and credit. These factors complicate access to non-traditional inputs, and to crop diversification and the

Patrick C. Wall is Agronomist, Global Maize Program, CIMMYT, P.O. Box MP 163, Mount Pleasant, Harare, Zimbabwe (E-mail: p.wall@cgiar.org).

[Haworth co-indexing entry note]: "Tailoring Conservation Agriculture to the Needs of Small Farmers in Developing Countries: An Analysis of Issues." Wall, Patrick C. Co-published simultaneously in *Journal of Crop Improvement* (Haworth Food & Agricultural Products Press, an imprint of The Haworth Press, Inc.) Vol. 19, No. 1/2 (#37/38), 2007, pp. 137-155; and: *Agricultural and Environmental Sustainability: Considerations for the Future* (ed: Manjit S. Kang) Haworth Food & Agricultural Products Press, an imprint of The Haworth Press, Inc., 2007, pp. 137-155. Single or multiple copies of this article are available for a fee from The Haworth Document Delivery Service [1-800-HAWORTH, 9:00 a.m. - 5:00 p.m. (EST). E-mail address: docdelivery@haworthpress.com].

establishment of crop rotations: an important practice to overcome pest and disease carryover in CA. Adequate equipment for direct seeding is a prerequisite for successful CA, but comparatively few resources have been dedicated to the development of direct-seeding equipment for low draught power conditions, given the relatively small profit margins associated with small equipment. All of these factors stress the need for the catalysis and development of innovation systems focusing on the development of CA in smallholder farming communities, and supporting the efforts of innovative farmers within these. Strategies for the successful adoption and management of CA practices need to address the issue of an enhanced knowledge base of individual farmers and the community. doi:10.1300/J411v19n01_07 *[Article copies available for a fee from The Haworth Document Delivery Service: 1-800-HAWORTH. E-mail address: <docdelivery@haworthpress.com> Website: <http://www.HaworthPress.com> © 2007 by The Haworth Press, Inc. All rights reserved.]*

KEYWORDS. Conservation agricultural systems, small farm characteristics, soil cover, crop residue, small-holder farmer

WHAT IS CONSERVATION AGRICULTURE?

Conservation agriculture (CA) comprises a suite of technologies which when used together are able to limit, arrest or revert many of the causes of unsustainable agricultural practices, such as soil erosion, soil organic matter decline, soil physical degradation and excessive pesticide and fuel use. The two principal characteristics of CA systems are minimal soil movement and continuous crop residue cover, supported by other components necessary to overcome problems associated with these practices, such as crop rotation and controlled traffic. The name "CA" has been widely adopted during the last 7-8 years to distinguish this more sustainable agriculture from the narrowly defined "conservation tillage"–taking the emphasis off the tillage component and adding the aspect of the full agricultural system rather than just a component of this.

Benefits of CA

The benefits of CA include both effects that start almost immediately and others that develop across time (Hamblin, 1987; Sayre, 1998; Derpsch, 1999). These benefits include:

Immediate Effects

- Increased water infiltration into the soil due to the protection of surface structure by the residues.
- Reduced water run-off and soil erosion due to the increased infiltration and the ponding effect of the residues.
- Reduced evaporation of moisture from the soil surface as the residues protect the surface from solar radiation.
- Better crop water balance, less frequent and intense moisture stress because of the increased infiltration and reduced evaporation.
- Reduced traction and labor requirements for land preparation, and thus savings in fuel and labor costs.

Medium-Term Effects

- Increased soil organic matter resulting in better soil structure, higher cation exchange capacity and nutrient availability, and greater water-holding capacity.
- Increased and more stable crop yields.
- Reduced production costs.
- Increased biological activity in both the soil and the aerial environment leading to more biological control of pests.

Farmer Adoption of CA

There has been increasingly rapid adoption of CA systems during the past 20 years: there was adoption of the systems prior to this, but adoption and advances were slower than they have been in more recent years. Derpsch (2005) has estimated, based on information from various sources, that there are approximately 95 million hectares worldwide of crops grown with zero tillage, one of the principal components of CA. Statistics on areas of particular cultural practices are difficult to obtain and it is likely that these figures mask very different system practices, some of which in fact might not classify as CA, such as those areas where the land is tilled for one crop and the following crop seeded without tillage.

Most of the area under zero tillage is on relatively large commercial farms using heavy tractors and equipment, whilst relatively little is on smallholdings, especially using manual or animal traction systems. There are, however, some important exceptions to this: in Brazil there is an estimated 200,000 hectares of permanent CA being practiced on small farms, especially in the States of Parana, Santa Catarina and Rio Grande do Sul; in the Indo-Gangetic Plains of India and Pakistan, there are an estimated 2 million hectares of wheat seeded without tillage, a large proportion of this by small farmers, although this is generally followed by an intensively tilled and puddled rice crop. In Ghana, by 2002 there were over 100,000 small farmers producing maize in CA systems, and pockets of adoption on small farms have been reported in several other countries. Reports from China are confusing and often conflicting, with some reports of millions of hectares while others suggesting that this is not the case.

The principles of CA appear to have extremely wide applicability: they are practiced on a wide range of soil types, from very heavy clays in Brazil to extremely sandy soil in Australia; from the Equator to at least 50°N latitude and with a wide range of crops including maize (*Zea mays* L.), wheat (*Triticum aestivum* L.), soybeans (*Glycine max* (L.) Merr), cotton (*Gossypium hirsutum* L.), sunflower (*Helianthus annuus* L.), rice (*Oryza sativa* L.), tobacco (*Nicotiana tabaccum* L.) and even root crops, such as potatoes (*Solanum tuberosum* L.) and cassava (*Manihot esculenta* Crantz), although obviously the harvest of these causes considerable soil movement. However, it has become evident that the actual techniques and technologies to implement the principles are very site-specific and require considerable local adaptation. Furthermore, although the principles of CA apply just as well to smallholders as to large, mechanized farmers, the techniques to apply them may be very different, given the different constraints of each group.

Problems of CA

One of the principal reasons farmers till the soil is to control weeds, and therefore, as tillage is obviated in CA, weed control becomes extremely important. In the first years of establishment of the CA system, weed control can be difficult and costly, but with good management, weed populations decline over time, due to the combined effects of seed bank depletion in the surface soil as seed is no longer incorporated by tillage, direct effects of residue cover on weed germination and the effects on seed viability of the increased populations of fauna and flora in the residue layer. Initially, however, weed control is generally achieved with the use of herbicides, especially glyphosate applied before crop emergence for total weed control. The greater requirement for herbicides in the initial stages of CA, compared with tilled systems, is a limitation of the technology.

Soil tillage aerates the soil, and the increased oxygen levels lead to rapid decomposition of soil organic matter in moist soil. Organic matter breakdown leads to the mineralization of the nitrogen in the residues, a fact exploited by farmers whose crops benefit from the greater levels of available nitrogen in the soil after intensive tillage. However, decomposition of both the surface residues and root mass is much slower in the undisturbed soil under CA, and thus there is not a flush of mineralized nitrogen for the young crop. Under certain circumstances, especially in soils with low organic matter content, greater levels (generally about 20 kg/ha) of nitrogen fertilizer need to be applied to CA crops to offset the reduction in available nitrogen in untilled soils. However, as soil organic matter levels increase under the new system, nitrogen mineralization increases and the need for this extra fertilization declines: after several years crops may need less applied N (and other nutrients) than in conventional systems to achieve the same or greater yields (Derpsch, 2005).

The CA system does not function well in soils with restricted drainage as the improved infiltration and reduced evaporation can exacerbate problems of water saturation and waterlogging, although the use of permanent raised beds (Sayre, 2004) may partially or completely overcome this limitation.

FACTORS AFFECTING THE ADOPTION OF CA ON SMALL FARMS

The characteristics and circumstances of smallholder farmers in the developing world are as diverse as the agro-ecologies and farming systems they manage. However, the following characteristics generally apply:

- Little access to financial capital.
- Prioritize production of family food needs, with sale of produce in excess of these requirements.
- Risk averse.
- Manage mixed crop/livestock systems.
- Limited land resources (although this is often not their primary limiting factor).
- Rely on manual labor, animal traction and/or small tractors for draught power, although they may contract service providers (with larger equipment) for some activities.

- Rely to a large degree on family members for hand labor.
- Have close community linkages with weaker links outside the community.
- Have less formal education than large-scale commercial farmers.
- Often are situated in marginal areas with respect to rainfall and topography.
- Often have precarious land tenure.

The following sections explore the effects that these characteristics have, or may have, on the adoption of CA systems.

The Importance of Knowledge

One of the most important changes necessary to adopt CA and zero tillage is a change in mind-set. The plow is often thought of as the symbol of agriculture, and making the leap to do away with tillage is difficult. Farmers generally undergo this change in mind-set relatively quickly when they experience, or are exposed to, the benefits of CA, but extension agents, researchers and university professors are slower to adapt to the new paradigm, especially those who have been preaching or enforcing the need for tillage. Although individual researchers and extension agents may be quick to see the benefits of CA and involve themselves in the new system, their institutions are generally much slower to evolve, and in several instances in developing countries research and extension directors have outlawed CA. Interestingly, in all such cases, there have still been advances in the technology as other agents supplant the public systems, but undoubtedly advances would have been far greater if the public institutions had been involved.

CA is a complex technology: it involves a complete change in the farming system, not simply the acquisition of a different crop seeder (as is often assumed). CA implies changes in weed control practices, in seeding dates, seeding times, crop residue management, crop rotations, harvest procedures and many other facets of the production system. As such, it is almost impossible for research and extension systems to develop appropriate "packages" that fit the circumstances of all farmers and farmer groups–the adaptation of the principles to local conditions requires large levels of farmer participation, and, indeed, in the more developed countries the advance of CA has been more farmer-driven than driven by research and extension institutions.

CA is more knowledge-intensive than input-intensive: success depends more on what the farmer does (management) than on the level of inputs he applies (Ekboir, 2002). This is one of the reasons for the far slower adoption of CA in the smallholder sector than on large farms. Whereas large farmers are often well connected to knowledge and information systems, this is generally not the case in small-farm communities. In these communities, information sharing within the community is often the primary source of new knowledge, and thus knowledge tends to be far more based on traditional concepts and practices than in the large farm sector where farmers tend to look outside their community for new knowledge to be able to give themselves an advantage in the market. Small farmers are not well connected to outside information systems: although they may own a radio, televisions are often not common, telephone systems are lacking, although this is improving with the advent of cell phones, and access to the Internet is almost non-existent. Outside technical information on agricultural practices generally comes from public extension agents who often themselves suffer from lack of access to new information, and may be slow to appreciate the feasibility and benefits of the CA system.

Peer pressure and community norms can also be important impediments to the adoption of practices that go completely against conventional wisdom. Even among large commercial farmers, this can be an important factor, and it is common for these farmers to tuck their first experiments on CA on back-fields where they are not open to the neighbors' view. There are many anecdotes about the trials of early adopters of zero tillage in the Indo-Gangetic Plains of South Asia: one of these tells of a young man who was approached by researchers to establish a demonstration plot of zero tillage wheat. He decided to accept, but when his father saw the stupid and lazy thing he had done, he banished him from the home and the village. However, when the incredulous farmer saw the beautiful germination of the crop without tillage, he sent out messengers to find his son and bring him back so that he could teach him and the villagers how to do this zero tillage (P. Hobbs, personal communication). Therefore, a concerted effort is needed to increase awareness among the whole community, not only of the problems caused by tillage, but also the benefits and feasibility of doing away with soil tillage. It is also important to develop groups of several farmers in the community to experiment with CA, so that they can provide mutual support to each other, especially in the face of doubt and ridicule.

Soil Cover with Crop Residues

Crop residue retention on the soil surface is the key to the success of CA systems, and the yield of crops direct-seeded into bare soil is often considerably lower than that of crops sown with conventional tillage practices (Wall, 1999; Sayre et al., 2001). However, smallholder farmers in developing countries generally manage intensive, mixed crop-livestock systems where animals are extremely important components of the system: they contribute to the food security of the household, provide for system diversification, generate cash, spread risk, recycle nutrients, provide draft power and transportation, and are important assets for investment and/or savings (de Haan et al., 1997). Crop residues are an important source of feed, especially for the larger ruminants, and, as such CA, and the need to leave crop residues on the soil surface, implies direct competition for a scarce resource. As pressure on the land increases, more arable land is dedicated to crops, intensifying the interactions and conflicts between crops and animals (Mueller et al., 2001).

In more marginal environments, crop productivity is lower and therefore crop residues are scarcer and competition for them greater. In areas with prolonged dry seasons, the demand for residues for feed is the greatest (McDowell, 1988; Sandford, 1989; Quiroz et al., 1997). Thus in the irrigated areas of the Indo-Gangetic plains where production levels are high and two or more crops per year are feasible, competition for residues between the needs of soil conservation and livestock feed is not very problematic, but in drier environments of northern and southern Africa, west Asia and parts of the Andes, this competition is extremely intense. The adoption of CA in these marginal environments will only advance when it can be demonstrated to farmers that leaving at least part of the residues on the soil surface gives a greater benefit to system productivity than feeding these to animals. That these increases in total system productivity are possible is evident in the results of Sayre et al. (2001): after several years of CA practices in central Mexico, productivity had increased sufficiently that more residues could be removed from the system for animal feed than in the conventional system, while still leaving sufficient for soil cover.

Sain and Barreto (1996) suggested that productivity levels might need to be raised to achieve sufficient levels of both ground cover and feed (Figure 1). At the present yield level (R_0), a farmer who uses half the stover production (A_0) for animal feed and the other half for ground cover (C), does not achieve the minimum requirement (M) for ground

FIGURE 1. System productivity under different management strategies and the assignment of crop residues to alternative uses: forage and ground cover. See text for description. Adapted from Sain and Barreto (1996).

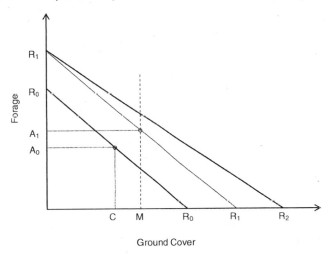

Ground Cover

cover. However, if productivity can be increased to R_1 (Sain and Barreto suggest by variety or improved fertilizer practices), then not only will there be more economic yield, but also the farmer can leave more residues, achieve the minimum amount of ground cover, and at the same time have more feed (A_1). However, the Sain and Barreto graph does not show the effect of residues on productivity: if mulch increases crop productivity as can be expected in rainfed and dryland situations, then allocating part of the crop residues to ground cover means not only that economic yield can be increased more by other agronomic interventions than it can without the mulch, but also that the effects on feed availability are greater (R_2).

Because of the importance of soil cover, successful establishment of the system is increasingly difficult in low productivity systems where enough crop residues cannot be produced to achieve adequate levels of ground cover for the following crop. There is still a lot of discussion on what is an "adequate" level of ground cover, but generally most authors suggest that 30% ground cover at seeding is necessary (e.g., Barber, 1996). This coincides with many studies on the effects of residue cover on soil erosion, where 30% ground cover reduces soil erosion by approximately 80% (Figure 2), resulting in the threshold ground cover amount in the definition of "conservation tillage." However, in systems

FIGURE 2. The relationship between relative erosion and ground cover (top graph) and between the amount of maize residues and ground cover (lower graph). Adapted from Erenstein (1997).

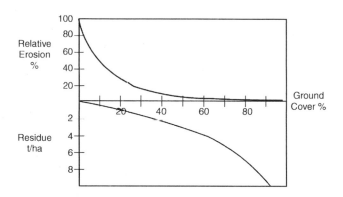

where erosion control is not the principal benefit obtained from CA systems, the optimum level of residues may be different. This question is still relevant in semi-arid areas where the principal benefit from CA is generally water savings and improved crop moisture balance.

The use of residues for animal feed also has a social component. In many regions, especially in rainfed areas, common grazing rights apply after crop harvest. Thus an individual farmer does not have exclusive rights to the residues on his land, and attempts to conserve them can lead to violent confrontation. For example, in central Mexico, Tripp et al. (1993) report several cases where farmers fenced their fields to keep residues, only to have the residues deliberately burned by irate neighbors. This complex issue can only be resolved by community awareness, and community involvement in the issue of land degradation, once again involving considerable investment in information sharing and knowledge development in rural communities.

Access to Inputs

CA, particularly in the initial years, may require higher levels of inputs, especially herbicides (or labor for manual or mechanical weeding) and possibly nitrogen fertilizer. For this reason, it is often thought that the system is not applicable to small-scale farmers, or that adoption will be limited. For instance, Rockstrom et al. (2001) state, "often herbicide use forms an integral part of promoted CT (conservation tillage) systems. Even though results generally show significant economic returns from minimum tillage systems using herbicides, the sustainability of promoting such a system among resource-poor farmers has been questioned, due primarily to problems of access and affordability." However, in a survey of farmer adoption of CA practices in Ghana, where the use of glyphosate herbicide is the base on which the system has been built, only 33% of farmers using zero-tillage with crop residue cover (of a total of 91 adopters interviewed) mentioned spending more money on inputs as a problem. However, all the farmers tried to reduce the dosage of the herbicide, and most farmers (70%) used less than the recommended 3 L/ha (Ekboir et al., 2001).

Although small-scale farmers have limited capital, and limited access to it, generally they are prepared to invest in inputs if the expected returns are sufficient and the risk of failure is low. Thus small farmers in the inter-Andean valleys of Bolivia will generally apply little or no inputs, other than seed and labor, to their wheat or barley crop, but will apply considerable amounts of fertilizer, manure and pesticides to their potato crop that has a higher market price. Potatoes are also grown on the most fertile soils where the risk of crop failure is low. One of the benefits of CA is generally a reduction in the risk of crop failure, related to an improved crop water balance. This reduction in risk is especially important for small farmers who have little savings to weather a bad harvest (Ekboir et al., 2001) and will tend to modify their willingness to invest in production inputs, including herbicides and N fertilizer.

However, the limited access to capital of small-scale farmers often means that access to input and output markets is limited–the volumes of sales or purchases are simply not sufficient to attract entrepreneurs to establish local businesses that stock the necessary inputs or buy the production, causing the problem of access referred to by Rockstrom et al. (2001) in the quote above. If CA is to spread in the small-scale agricultural sector, issues of access to relevant inputs at reasonable prices need to be addressed, and innovative methods to stimulate the development of input and output markets in smallholder farming areas need to be established. These may include programs to assume part or all of the downside risk to retailers, ensuring sales or absorbing un-sold inputs, and subsidizing local output markets.

Another important aspect of this problem is that of equipment: without functional equipment for direct seeding of crops it is impossible to properly test, and, more importantly, demonstrate the benefits of CA. While large machinery manufacturers have seen the beneficial financial prospects of developing and manufacturing zero tillage seeders for large-scale farmers, in most places similar efforts have not focused on equipment for small-scale farmers. In Brazil, public research efforts were dedicated to overcome this problem in the late 1980s and early 1990s, producing zero tillage animal-traction seeders for large seeded crops. Once CA started to expand in the smallholder sector, several companies began to develop and manufacture equipment for this sector, which then spurred continued expansion of CA on small farms. In Pakistan, CIMMYT agronomists supported the development and manufacture of zero tillage small-grain seed drills for use with small tractors. Once these drills were locally produced, it stimulated more farmer experimentation with CA, and then fed the growth of zero tillage wheat in the western Indo-Gangetic Plains. Similar efforts by university researchers in north-west India also resulted in adapted equipment, which stimulated early adoption of zero tillage, which in turn stimulated expansion of local manufacturing of small-scale equipment. By 2002, there were at least 11 manufacturers of zero tillage small grain drills in north-western India.

Labor Use and Labor Savings

One of the most important benefits of CA for small farmers is the reduction in labor requirements for tillage (e.g., Sorrenson et al., 1998). However, if only manual weeding is practiced, then labor requirements may increase, making the system unattractive to farmers. Muliokela et al. (2001) state that zero tillage systems based on mulching, planting and weeding by hand have not met with success in Zambia and Zimbabwe due to excessively high labor requirements.

In many places, the non-availability of family labor is a major limiting factor, often defining the amount of land that is farmed. In these circumstances, the amount of food that can be produced per day of labor may be more important than the amount of food produced per unit of land area. Thus in Ghana, Ekboir et al. (2001) found that the most important impact of zero tillage on small-scale farmers was the reduction in the amount of labor used per unit of agricultural output. In this system, farmers control weeds with glyphosate before crop emergence, and then control further weed emergence and growth manually. The labor requirements for weeding, compared with the conventional production system, were reduced by 50% (8.8 person days/ha to 4.3 person days/ha), allowing farmers extra time to undertake other activities. In the survey of small-scale farmers in Ghana reported by Ekboir et al. (2001),

99% of the farmers who were using zero tillage said that zero tillage was less physically demanding than the conventional production system, and also that it reduced labor requirements at critical times, making labor management easier.

Timeliness of Seeding

One of the most important drivers of zero tillage wheat in the Indo-Gangetic Plains has been the possibility of seeding the wheat crop earlier. Research had shown that wheat yield was reduced by 1-1.5% for each day that seeding was delayed after November 30 (Hobbs et al., 1997). After the harvest of the rice crop, tillage required several days, delaying the date of seeding of the wheat. However, without tillage the wheat crop could be seeded almost immediately after rice harvest, ensuring more timely seeding. Another important advantage of zero-tilled wheat in the western IGP has been the effect of earlier sowing on the most important weed in the wheat phase of the system–*Phalaris minor*. Earlier seeding allows the wheat crop to establish before temperatures drop sufficiently for *Phalaris* germination, and the weed then cannot compete with the crop. An analysis by the Center for International Economics, Canberra, Australia, showed that the advantage to agriculture in the State of Haryana, India, due to the savings in herbicide and improved weed control was worth US$ 238 million across the next 30 years.

In areas with a long dry season, animals are weak at the end of the dry season due to limited feed availability, making soil tillage difficult and time consuming and leading to late seeding of the crop (Rockstrom et al., 2001). Many small-scale farmers rely on renting draught animals or tractors for tillage, but seldom do they manage to get these during the period of peak demand for tillage at the start of the rainy season, and again, seeding of the crop is delayed. This offers an opportunity for zero tillage and CA in these situations; indeed much of the interest in CA expressed by farmers in eastern and southern Africa focuses on the potential for timely crop seeding.

Development of Innovation Systems

The rapid spread of CA among large, commercial farmers in the Americas and Australia has largely been a phenomenon driven by farmers and farmer organizations, catalyzed initially by commercial interests of input suppliers (Ekboir, 2002). Innovation systems developed around the new technology, with multiple players, including farmers, researchers, extension agents, input suppliers, machinery manufacturers and many others engaging in a complicated system of information flow in multiple directions. In his review of case studies of adoption of zero tillage, Ekboir (2002) states, "although the development of no-till packages and their adoption by small-scale farmers followed different paths than for large-scale farmers, the paths shared one important common feature: all successful programs resulted from networks that worked with participatory research approaches."

Commonly, research and extension systems in the developing countries follow a linear model of agricultural research and extension and knowledge flow is uni-directional (Figure 3). In this model, researchers develop technological options that are transferred to farmers through extension agents (Ekboir, 2002), and farmers do not participate in technology development. While these systems may function well for simple technological changes, they are inadequate for the development and adoption of complex technologies, such as CA (Wall et al., 2002). What is required is for interested agents to catalyze innovation systems, based on the activities of innovative farmers, to help overcome problems observed in farmer experiments, and help achieve the potential of the system. These highly participatory innovation systems do not develop without a catalyst: an individual or organization with a specific interest in advancing the technology and its adoption. In the early years of CA development, this role was played by input manufacturers (specifically ICI trying to develop a market for their paraquat herbicide), but in different places has included farmers, and national and international researchers. There is an urgent need to support CA innovation systems in smallholder faming communities, and especially to support the individuals who have the foresight and tenacity to catalyze these systems.

Political Aspects of CA Adoption

Land tenure is an important factor in the decision process with respect to technologies that may only yield benefits after several years: if there is a risk that today's investments (in labor or other inputs) will not benefit the investor in the future, it is unlikely that farmers will commit themselves to the investment. Farmers in smallholder communities generally have less clear land tenure than larger, commercial farmers, although there are large national and regional differences in this respect. Unless farmers have clear ownership of the land, they are unlikely to adopt CA: non-ownership categories of land use, such as leasing, rent-

FIGURE 3. A representation of the linear model of technology and knowledge development in agriculture. Based on Ekboir, 2002.

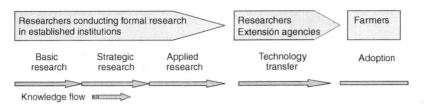

ing, or share-cropping lead to farmers having little motivation to holistically manage crops, livestock and natural resources (Thiesenhusen, 1994).

The adoption of any new technology implies a cost, and the investment in acquiring new knowledge of a complex system, such as CA, may be high for a risk-averse smallholder farmer. Given the important off-farm effects of the adoption of technologies that reduce or revert land degradation, there is a good argument for other beneficiaries to support CA adoption and to compensate farmers for their efforts in land stewardship. This support may take many forms, from subsidies on CA equipment (possibly counterbalanced by higher taxes on tillage equipment), direct payments to farmers for environmental services, such as those envisaged in payments for carbon credits, or simply credit schemes for farmers linked to the adoption of conservation practices. However, to achieve this type of support, there is a need for education of policy makers in the costs of land degradation, the existence of technologies to overcome this and the expected benefits that would accrue to the region or nation from the widespread adoption of CA and related technologies.

CONCLUSION

The main issues involved in achieving widespread adoption of CA in smallholder systems described in the sections above are:

- Mind-set
- Knowledge of the CA system
- Residue retention, and competition for scarce residues
- Physical and financial access to inputs

- Availability of adapted equipment
- Capacity building among farmers, researchers and extension agents.
- Development of innovation systems around CA
- Land tenure
- Support to farmers for environmental services

These aspects need to be incorporated in a holistic manner into efforts to stimulate and support the adoption of CA among smallholder farmers.

Facilitating the Widespread Adoption of CA in Smallholder Maize-Based Systems in Eastern and Southern Africa

Driven by an analysis of the issues involved in achieving widespread adoption of CA by smallholder farmers in South America, the Indo-Gangetic Plains and Ghana, CIMMYT has recently started a project to evaluate a holistic approach to the problem of CA adoption in four countries in Eastern and Southern Africa. To date, there has not been widespread adoption of CA (with residue cover) in dryland areas where crop residues are used for animal feed (Ekboir, 2002). This project includes activities to clarify and overcome many, if not all, of the limitations to adoption of CA by smallholder farmers discussed above, including:

- Community awareness activities, aimed at increasing community knowledge on the issues of soil and land degradation, and avenues and attitudes to overcome these.
- Farmer training in CA techniques.
- Demonstration plots on farmer fields, initially based on "best-bet" combinations of technologies, but incorporating new technologies and techniques as these are developed.
- Organization of groups of interested farmers focusing on the demonstration plots.
- Farmer experiments: areas where farmers try the technology on their own fields. These are the focal point of the project.
- Facilitation of farmer-to-farmer exchange of information, both within the community and outside the community, concentrating on areas with similar biophysical and socioeconomic characteristics. Farmer experiments play a key role in this effort.
- Researcher and extension agent training in CA principles and practices.

- Participatory evaluation and modification of equipment for seeding and spraying in animal traction and manual systems, and stimulation of local manufacture of adapted equipment.
- On-farm and on-station research trials to develop solutions to problems observed in demonstration plots and farmer experiments.
- Long-term trials under experiment station conditions to monitor the effects of CA practices on crop productivity and soil physical, chemical and biological characteristics. Data from these trials will allow for the modeling of benefits at different scales, and should provide important information for policy makers.

This project is focused on a limited number of communities (3-4) in each of Malawi, Tanzania, Zambia, and Zimbabwe. Baseline surveys have been conducted in all communities, and further surveys and case studies will be conducted at the end of the project to assess the effectiveness of the approach to promote adoption of locally adapted CA practices, and their effects on farm family income and livelihoods.

REFERENCES

Barber, R.G. 1996. Linking the production and use of dry-season fodder to improved and conservation practices in El Salvador. Proyecto CENTA-FAO, AG:GCP/ELS/004/NET. Documento de campo No. 8.

de Haan, C., H. Steinfeld and H. Blackburn. 1997. Livestock and the environment: Finding the balance. European Commission, Directorate-General for Development. Brussels, Belgium. 115p.

Derpsch, R. 1999. Expansión mundial de la SD y avances tecnológicos. In: Proceedings of the 7th National Congress of AAPRESID, 18-20 August, 1999. Mar del Plata, Argentina.

Derpsch, R. 2005. The extent of conservation agriculture adoption worldwide: Implications and impact. In: "III World Congress on Conservation Agriculture: Linking Production, Livelihoods and Conservation," Nairobi, October 3-7, 2005. (CD).

Ekboir J., K. Boa and A.A. Dankyi. 2001. The impact of no-till in Ghana. In: Conservation Agriculture: A Worldwide Challenge. García-Torres, L., Benites, J., Martínez-Vilela, A. (eds.). ECAF/FAO, Córdoba, Spain. Vol II, pp. 757-764.

Ekboir J. 2002. Part 1. Developing no-till packages for small farmers. In: Ekboir J. (ed.) 2002. CIMMYT 2000-2001 World Wheat Overview and Outlook: Developing No-Till Packages for Small-Scale Farmers. Mexico, DF: CIMMYT. pp 1-38.

Erenstein, O. 1997. ¿Labranza de conservación o conservación de residuos? Una evaluación del manejo de los residuos en México. NRG Reprint Series 97-02. México, D.F.: CIMMYT. 10 p.

Hamblin, A. 1987. The effect of tillage on soil physical conditions. In: P.S. Cornish and J.E. Pratley (eds.) Tillage: New Directions in Australian Agriculture. Australian Society of Agronomy. pp. 128-170.

Hobbs, P.R., G.S. Giri and P. Grace. 1997. Reduced and zero tillage options for the establishment of wheat after rice in South Asia. RWC Paper No. 2. Mexico, D.F.: Rice-Wheat Consortium for the Indo-Gangetic Plains and CIMMYT.

McDowell, R.E. 1988. Importance of crop residues for feeding livestock in smallholder farming systems. In: J.D. Reed, B.S. Capper and P.J.H. Neute (eds.) Plant Breeding and the Nutritive Value of Crop Residues. Proceedings of a workshop. ILCA, Addis Ababa (Ethiopia), pp. 3-27.

Mueller, J.P., D.A. Pezo, J. Benites and N.P. Schlaepfer. 2001. Conflicts between conservation agriculture and livestock over the utilization of crop residues. In: Conservation Agriculture: A Worldwide Challenge. García-Torres, L., Benites, J., Martínez-Vilela, A. (eds.). ECAF/FAO, Córdoba, Spain. Vol I, pp. 211-225.

Muliokela S.W., W.B. Hoogmoed, P. Stevens and H. Dibbits. 2001. Constraints and possibilities for conservation farming in Zambia. In: Conservation Agriculture: A Worldwide Challenge. García-Torres, L., Benites, J., Martínez-Vilela, A. (eds.). ECAF/FAO, Córdoba, Spain. Vol II, pp. 61-65.

Quiroz, R.A., D.A. Pezo, D.H. Rearte and F. San Martin. 1997. Dynamics of feed resources in mixed farming systems of Latin America. In: C. Renard (ed.) Crop Residues in Sustainable Mixed Crop/Livestock Systems. CABI, Wallingford, UK. pp. 149-180.

Rockstrom, J., P. Kaumbutho, P. Mwalley and M. Temesgen. 2001. Conservation farming among small-holder farmers in E. Africa: Adapting and adopting innovative land management options. In: Conservation Agriculture: A Worldwide Challenge. García-Torres, L., Benites, J., Martínez-Vilela, A. (eds.). First World Congress on Conservation Agriculture, Volume 1: Keynote Contributions 39:363-374, FAO, Rome, Italy. ECAF/FAO, Córdoba, Spain.

Sain, G.E. and H.J. Barreto. 1996. The adoption of soil conservation technology in El Salvador. Linking productivity and conservation. J Soil and Water Cons. 51(4):313-321.

Sandford, S.G. 1989. Crop residue/livestock relationships. In: Soil, Crop and Water Management in the Sudano-Sahelian Zone. Proceedings of an Internattional Workshop. Jan. 11-16, 1987. ICRISAT Sahelian Center, Niamey, Niger. ICRISAT, Patancheru, pp.169-182.

Sayre, K.D. 1998. Ensuring the use of sustainable crop management strategies by small wheat farmers in the 21st century. Wheat Special Report No. 48. Mexico, D.F. CIMMYT.

Sayre, K.D. 2004. Raised-bed cultivation In: R. Lal (ed.) Encyclopaedia of Soil Science. Marcel Dekker Inc. (eBook Site Online Publication 04/03/2004).

Sayre, K.D., M. Mezzalama and M. Martinez. 2001. Tillage, crop rotation and crop residue management effects on maize and wheat production for rainfed conditions in the altiplano of central Mexico. In: Conservation Agriculture: A Worldwide Challenge. García-Torres, L., Benites, J., Martínez-Vilela, A. (eds.). ECAF/FAO, Córdoba, Spain. Vol II, pp. 575-580.

Sorrenson, W.J., Duarte, C. and López Portillo, J. 1998. Economics of no-till compared to conventional cultivation systems on small farms in Paraguay, policy and investment implications. Report Soil Conservation Project MAG–GTZ, August 1998.

Thiesenhusen, W.C. 1994. The relation between land tenure and deforestation in Latin America. In: J.E. Oman and D. Pezo (eds). Proceedings, Symposium/Workshop. "Livestock and Natural resources in Central America: Strategies for Sustainability." San Jose, Costa Rica, 7-12 October, 1991. CATIE, Turialba, Costa Rica. pp. 243-256.

Tripp, R., D. Buckles, M. van Nieuwkoop and L. Harrington. 1993. Land classification, land economics and technical change. Akward issues in farmer adoption of land-conserving technologies. Seminar presented in "Seminario/Taller Internacional para la Definición de una Metodología de Evaluación de Tierras para una Agricultura Sostenible en México," El Batan, México, 10-13 August, 1993.

Wall, P.C. 1999. Experiences with crop residue cover and direct seeding in the Bolivian highlands. Mountain Research and Development 19:4, 313-317.

Wall, P.C., J.M. Ekboir and P.R. Hobbs. 2002, Institutional aspects of Conservation Agriculture. Paper presented at the International Workshop on Conservation Agriculture for Sustainable Wheat Production in Rotation with Cotton in Limited Water Resource Areas, Tashkent, Uzbekistan, October 13-18, 2002.

doi:10.1300/J411v19n01_07

Improved Cowpea-Cereals-Based Cropping Systems for Household Food Security and Poverty Reduction in West Africa

B. B. Singh
H. Ajeigbe

SUMMARY. Food production in West Africa has not been keeping pace with the population growth because the bulk of the agriculture in this region is still based on traditional inter-cropping systems with little or no application of fertilizers and chemicals. The average use of fertilizers in West Africa is less than 10 kg/ha/year. This leads to a negative balance of nutrients in the soil and continuous decline in crop yields, which perpetuates malnutrition, hunger and poverty through the vicious circle of 'low input–low production–low income' and food insecurity. How to reverse this trend is one of the major challenges of agricultural research in this region. The International Institute of Tropical Agriculture (IITA), in collaboration with relevant national, regional and international partners, has developed an appropriate model that seems to hold great promise for

---

B. B. Singh is Cowpea Breeder and H. Ajeigbe is Officer-in-Charge, International Institute of Tropical Agriculture (IITA), P.M.B. 3112, Kano, Nigeria.

Address correspondence to: B. B. Singh at the above address (E-mail: b.b.singh@cgiar.org).

[Haworth co-indexing entry note]: "Improved Cowpea-Cereals-Based Cropping Systems for Household Food Security and Poverty Reduction in West Africa." Singh, B. B., and H. Ajeigbe. Co-published simultaneously in *Journal of Crop Improvement* (Haworth Food & Agricultural Products Press, an imprint of The Haworth Press, Inc.) Vol. 19, No. 1/2 (#37/38), 2007, pp. 157-172; and: *Agricultural and Environmental Sustainability: Considerations for the Future* (ed: Manjit S. Kang) Haworth Food & Agricultural Products Press, an imprint of The Haworth Press, Inc., 2007, pp. 157-172. Single or multiple copies of this article are available for a fee from The Haworth Document Delivery Service [1-800-HAWORTH, 9:00 a.m. - 5:00 p.m. (EST). E-mail address: docdelivery@haworthpress.com].

Available online at http://jcrip.haworthpress.com
doi:10.1300/J411v19n01_08

increasing food production in West Africa without affecting the environment and degrading the soils. This model involves a holistic combination of new, more productive dual purpose and resilient cultivars of cowpea [*Vigna unguiculata* (L.) Walp.], maize (*Zea mays* L.), sorghum [*Sorghum bicolor* (L.) Moench] and millet (*Pennisetum glaucum* and other species) in a strip-cropping pattern with a minimum and selective application of fertilizers and pesticides, feeding of crop residues to small ruminants in permanent enclosures on the home compound and returning of manure to the field. Based on this model, two 'best bet' options have become popular with farmers in northern Nigeria. These are: (1) an improved strip-cropping system involving two rows of a densely planted, improved sorghum variety: four rows of a densely planted, improved medium-maturity cowpea variety in the Sudan savanna where the rainfall is about 600 mm and (2) an improved strip-cropping system involving two rows of a densely planted, improved maize variety: four rows of densely planted double cropping of an improved 60-day cowpea in the northern Guinea savanna where the rainfall is about 1000 mm. The two-third cowpea and one-third cereal combination minimizes fertilizer use and maximizes profit because of the higher prices of cowpea grain and fodder and at the same time it leaves positive residual soil nitrogen balance and reduces *Striga hermonthica* seed bank, both of which benefit the cereal crops. This combination is also appropriate in view of the global surplus of cereals and global deficit of legumes. The on-station and on-farm evaluation of these systems covering several states and more than 2000 farmers in northern Nigeria, with the financial support from USAID, Gatsby Foundation and DFID, has shown over 300% increase in productivity, enhanced income generation and improved livelihoods of the farm families. doi:10.1300/J411v19n01_08 *[Article copies available for a fee from The Haworth Document Delivery Service: 1-800-HAWORTH. E-mail address: <docdelivery@haworthpress.com> Website: <http://www.HaworthPress.com> © 2007 by The Haworth Press, Inc. All rights reserved.]*

KEYWORDS. Cowpea, cereals, intercropping, planting pattern, strip-cropping

INTRODUCTION

The rapid increase in population and consequent pressure for food are driving agriculture towards greater intensification in West Africa

(Sanginga et al., 2003). The long fallow periods have not only diminished but also agriculture has now been pushed onto marginal lands, leaving little or no scope for further expansion in the cultivated area. This is more pronounced in the dry savannas of West Africa where rainfall is low and soils are predominantly sandy with low organic matter, low phosphorus and poor water holding capacity. Also, agriculture in this region is still based on traditional inter-cropping of cereals like maize (*Zea mays* L.), sorghum [*Sorghum bicolor* (L.) Moench] and pearl millet (*Pennisetum glaucum*), with a cowpea arrangement involving 1-row cereal–1-row cowpea [*Vigna unguiculata* (L.) Walp.] with a low plant density and little or no application of fertlizers and chemicals (Norman, 1974; Baker and Norman, 1975; Edwards, 1993; Mortimore et al., 1997; Henriet et al., 1997). The FAO estimates (FAO STAT- 2004) indicate that West Africa uses about 8 kg fertilizer/ha annually compared with 100 to 400 kg/ha in other countries (Table 1). Such practices lead to decreased soil organic matter, increased populations of chronic parasitic weeds (e.g., *Striga* spp.), reduced soil biological diversity and enhanced erosion risk. This, in turn, leads to a negative balance of nutrients in the soil and continuous decline in crop yields. Recent estimates indicate that the annual nutrient losses exceed 26 kg of N, 3 kg of P, and 19 kg K per hectare (Sanginga et al., 2003). Consequently, in this system, the cereal yields are low due to low fertility and cowpea yields are low due to shading from cereals throughout the growing period and thus the overall grain as well as fodder productivity of both cereals and cowpea is drastically reduced. Even though cowpea occupies 50% of the land area under intercropping, its yield is usually less than 25% of the sole crop and farmers obtain less than 1 ton of total food per hectare (van Ek et al., 1997), which perpetuates poverty through the vicious circle of 'low input–low production–low income' and consequent food insecurity. Presently, over 70% of the population in West Africa lives below the poverty line, spending less than one dollar a day. The widely reported food crisis in 2005 in Niger Republic is an example of this problem. How to help them and reverse this trend?

It is a known fact that without adequate inputs, yields cannot be increased, but at the same time large quantities of fertilizers and chemicals are neither available nor desirable in West Africa where soils have very low organic matter. Thus, the major challenge of agricultural research in this region is to maximize the benefits of small amounts of purchased inputs like fertilizers and pesticides. In view of the limited availability of fertilizers and low soil organic matter in West Africa, use of a greater proportion of improved varieties of fast-growing and nitrogen-fixing le-

TABLE 1. Total fertilizer use and average crop yields in different countries, 2001.*

Country	Fertilizer X 10³ t	Arable land X 10³	Fertilizer kg/ha	Yield t/ha Cereals	Pulses
Nigeria	221	28,200	7.80	1.04	0.40
Niger	5	4,994	0.98	0.40	0.12
West Africa	472	56,465	8.35	0.92	0.37
China	35,375	124,140	284.00	4.80	1.35
India	17,359	161,750	107.00	2.40	0.53
Netherlands	401	914	439.00	7.20	4.40
USA	19,614	177,954	110.00	5.90	1.83
World	137,729	1,369,110	100.00	3.10	0.79

*FAOSTAT: April 2004.

gumes like cowpea in slightly modified cropping systems to minimize shading and competition and recycling the crop residues as manure has been thought to be a possible strategy to partly address this problem.

Cowpea is an important source of nutritious food and fodder in the semi-arid tropics, especially in West Africa. The traditional varieties grown as intercrop with cereals have 'spreading' growth habit and take more than 130 days to mature with less than 1 t ha^{-1} yield potential. Through systematic breeding, improved grain-type and dual purpose (grain + fodder cowpea) varieties have been developed with erect and semi-erect growth habit and 60- to 75-day maturity, some of which yield up to 2.5 t ha^{-1} grain and fodder with 25% to 29% grain protein and 15% to 18% protein in the haulms (Singh, 1994; Singh et al., 1997; Singh et al., 2003). These varieties also fit well in the niches of existing cereal-based cropping systems and contribute to soil fertility and reduce the seed bank of *Striga* species, thereby enhancing cereal productivity (Terao et al., 1997; Mortimore et al., 1997; Singh and Ajeigbe, 2002). This has made it possible to intensify agriculture in West Africa and make more efficient use of resources and enhance crop-livestock integration. This article describes on-station and on-farm trials of new improved intensive cropping systems involving 'maize-double cowpea,' 'sorghum-cowpea,' and 'millet-cowpea' strip cropping with up to 100% to 300% increase in total productivity and gross income compared with the traditional intercropping systems.

MATERIALS AND METHODS

Several on-station experiments were conducted to evaluate different intercropping systems involving selected varieties of millet, sorghum and cowpea. The traditional intercropping involving 1 row of cereal and 1 row of cowpea causes severe shading by fast-growing, tall plants of millet and sorghum and reduced growth of cowpea. Therefore, wider strips involving 1 row cereals:2 rows cowpea; 2 rows cereals:2 rows cowpea; 2 rows cereals:4 rows cowpea and sole crops of cereals and cowpea were evaluated along with 1 cereal:1 cowpea to assess and compare the overall productivity and economic returns from the various systems. In all experiments, improved varieties of cowpea and local varieties of millet and sorghum were used. Before planting, a basal dose of compound fertilizer at the rate of 100 kg NPK (15:15:15)/ha and one ton of farmyard manure was applied to the experimental plots. Planting was done on ridges that were 75 cm apart. Each plot consisted of 6 rows that were 9 m long. The number of cereals or cowpea rows depended upon the intercrop treatment. The cereal rows were top dressed twice with urea 3 weeks after planting and 6 weeks after planting at the total rate of 50 kg urea (46%N) per ha and cowpeas were sprayed with Cypermethrin at the rate of 1 litre/ha at flowering and at podding stages. Net plots depended upon the intercrop treatments, but yields of all the component crops were expressed on per ha basis.

To validate the results of these on-station trials, several on-farm trials were also conducted. A total of 15 farmers in Kaduna and Kano states (Nigeria) planted the improved 'maize-double cowpea' and 'sorghum-cowpea' strip systems. The maize-double cowpea system involved 2 rows of maize variety 'Across 97' and 4 rows of 60-day cowpea variety 'IT93K-452-1' on 0.4 ha to 1.0 ha land. A basal dose of 100 kg NPK (15:15:15)/ha was applied before planting. The first cowpea was planted at the same time as maize in the first week of June. Maize was topdressed with 50 kg urea/ha and cowpea was sprayed twice with Cypermethrin to control insect pests. The first cowpea matured in about 60 days and was harvested in the first week of August when the maize was still green and had not tasselled yet. The cowpea grains were used for food and cash income and the cowpea residues were incorporated into the soil to provide nutrients to the standing maize crop and incoming second cowpea crop. The second cowpea variety 'IT89KD-288' was planted immediately after the harvest of the first cowpea (between August 10 and 20) in all of the rows, including maize rows. The maize was hand-harvested in October and the second cowpea was harvested in November. The first cowpea provided food, green manure/green forage and cash in the lean period (mid-rainy season), whereas the second cowpea and maize provided food, fodder and cash.

The improved strip cropping system in Kano State involved 2 rows of sorghum:4 rows of single crop of a medium-maturity cowpea variety 'IT90K-277-2' because of the relatively low rainfall in this region. A total of 15 farmers planted between 0.4 ha and 1.0 ha plots of the improved sorghum-cowpea system. Both the cowpea and sorghum were densely planted. The purchased inputs included 100 kg of NPK (15:15:15) as a basal dose, 50 kg of urea for top-dressing sorghum and 1 L of Cypermethrin insecticide to spray cowpea at flowering and podding stages to protect from insects.

Following encouraging results from these trials, this system is now being widely evaluated involving a large number of farmers in several states in northern Nigeria through financial support from USAID, Gatsby Foundation and DFID.

RESULTS

The results of on-station trials involving millet and cowpea are presented in Table 2. Considering the grain and fodder yields of cowpea and millet in different systems and placing an economic value to them, the 2 millet-4 cowpea and sole cowpea systems gave the highest gross returns and sole millet the lowest. It was also evident that the cowpea yield under the 1 row millet-1 row cowpea system was much lower than expected due to a severe competition with millet. The results from the sorghum-cowpea intercropping trials were also similar to those from the millet-cowpea intercrops (Table 3). The highest gross economic returns were obtained from the sole crop of cowpea and the 2 sorghum-4 cowpea system. Again, the yield of cowpea was much lower than expected from the 1 sorghum-1 cowpea intercrops.

The gross income per ha from the 2 rows millet-4 rows cowpea system was Naira 55,514 (about US$411) compared with Naira 38,647 (about US$286) for the 1 row millet-1 row cowpea. Similarly, the gross income from the 2 rows sorghum-4 rows cowpea system was 60,135 Naira (about US$445) compared with Naira 39,156 (about US$290) for the 1 row sorghum-1 row cowpea system. Also, the grain and fodder yields of cowpea were higher in the 2-row:4-row systems, which en-

TABLE 2. Cowpea and millet productivity in different planting patterns, 2001.

| Pattern | Yield kg/ha | | | | LER | LER | Total | Naira* |
| | Cowpea | | Millet | | Cowpea | Millet | LER | Value |
	Grain	Fodder	Grain	Fodder				
1M:1C	340	580	1530	3747	.26	.72	.98	38647a
1M:2C	485	709	1049	2322	.36	.49	.85	37557a
2M:2C	390	576	1433	3230	.29	.68	.97	38684a
2M:4C	841	1652	898	2087	.63	.42	1.05	66610b
Sole millet	-	-	2121	6292	-			31742a
Sole cowpea	1331	1848			-			62700b

*Based on 2001 prices.
Cowpea grain @ N35/kg, cowpea fodder @N8/kg, millet and sorghum grain @12/kg, millet and sorghum fodder @ N1/kg.
Conversion rate 1 US dollar = 100 naira; M = millet, C = cowpea.
a, b = the same letters do not differ significantly.

TABLE 3. Cowpea and sorghum productivity in different planting patterns, 2001.

| Pattern | Yield kg/ha | | | | LER | LER | Total | Naira* |
| | Cowpea | | Sorghum | | Cowpea | Sorghum | LER | Value |
	Grain	Fodder	Grain	Fodder				
1S:1C	343	946	1187	5330	.26	.04	1.10	39150a
1S:2C	432	1417	842	5137	.32	.60	.92	41697a
2S:2C	378	1065	965	5317	.28	.68	.96	38647a
2S:4C	914	1652	490	3497	.67	.35	1.02	60135b
Sole sorghum			1411	1042	-			27174c
Sole cowpea	1331	1848			-			61369b

*Based on 2001 prices.
Cowpea grain @ N35/kg, cowpea fodder @N8/kg, millet and sorghum grain @12/kg, millet and sorghum fodder @ N1/kg.
Conversion rate 1 US dollar = 100 naira; S = sorghum, C = cowpea.
a, b = the same letters do not differ significantly.

sured nutritious food for participating farmers' families and nutritious fodder for the livestock.

The component grain and fodder yields and gross returns from on-farm trials in Kaduna State (Nigeria) involving 2 rows maize-4 rows double cowpea are given in Table 4. The grain and fodder yields of cowpea and maize represent 13 participating farmers (2 farmers could not complete the trials). The best farmer, Mr. Muhammad Shehu, had a gross income of Naira 212,280 per ha, while the mean value for all

TABLE 4. Grain and stover yields (kg/ha) obtained by farmers participating in the improved strip-cropping system at Ungwa Namama, Kaduna State, 2003.*

Name	First Cowpea Grain	First Cowpea Fodder	Maize Grain	Maize Fodder	Second Cowpea Grain	Second Cowpea Fodder	Naira Values* Grain	Naira Values* Fodder	Naira Values* Total
Muhammadu Shehu	700	1319	2355	2775	1754	515	179224	33056	212280
Abdulsalam Shehu	733	1465	1578	1598	1376	653	143326	34965	178291
Adamu Rabiu	915	1492	1111	1332	1355	684	140208	35302	175510
Alh. Abdu Aliyu	789	1372	1289	1465	1467	497	143697	30969	174666
Aliyu Abdusalamu	733	1410	1200	1887	1288	746	129863	36108	165971
Haruna Yusuf	789	944	1600	2109	955	799	125587	30359	155945
Shafiu Shehu	767	1254	1311	1665	1111	515	125343	29870	155213
Shafiu Rabiu	733	986	1200	1665	1354	1284	104363	37375	141738
Alh. Isah	667	1199	1289	1221	699	1021	99210	35742	134952
Ibrahim Shehu	533	1299	1422	1288	511	995	86347	36974	123321
Jibrin Abdulsalamu	869	1185	711	1887	466	620	83805	30856	114661
Dahiru Abdulsalam	578	719	733	1221	599	844	76434	25885	102319
Abdu Jibrin	655	744	700	1452	622	844	63883	26713	90596
Mean	728	1184	1327	1659	1043	770	115484	32629	148113
Control (8)	940	401	1913	1917	-	-	94792	9849	104641

* Based on 2003 prices: cowpea grain = N50/kg, cowpea fodder = N15/kg, maize grain = N24/kg, maize fodder = N2/kg.

farmers was Naira 148,113 per ha. This compared very well with the mean income of Naira 104,641 from the improved control systems prevalent in the region. A mean total food grain yield of 3,098 kg/ha was obtained, which comprised 1,771 kg of cowpea and 1,327 kg of maize, and a mean total fodder yield of 3,613 kg/ha was made up of 1,954 kg of cowpea fodder and 1,659 kg of maize stover. The income from grain alone was Naira 115,484. After paying Naira 7,500 for the purchased inputs, the participating farmers made a good profit. All the participating farmers were very happy with the economic returns and more farmers have adopted the system.

The grain and fodder yields of cowpea and sorghum for 14 participating farmers in Kano State, Nigeria (one farmer could not complete the trial) are presented in Table 5. The best farmer, Mr. Danyaro Kawu, had a gross income of Naira 176,490 per ha, with the mean gross income being Naira 133,926 for all farmers. This was substantially more than the income from the control systems prevalent in the area (Naira 35,944 to 57,135 per ha). The cost of purchased inputs was about Naira 7,000;

TABLE 5. Grain and stover yields (kg/ha) of farmers participating in the improved strip cropping system in Wudil, Kano State, 2003.

Farmer	Grain	Fodder	Grain	Fodder	Grain	Fodder	Total
Danyaro Kawu	2264	4210	Not	Planted	113220	63270	176490
Alh. Ilu Tela	1376	2753	1376	6216	101854	53724	155578
Mall. Mamuda	1154	3996	1066	4440	83294	68820	152114
Baffa A. Abdu	622	4174	1598	7548	69442	77700	147142
Alh. Sabo Mail Wake	1221	2442	1576	5550	98870	47730	146600
Alh. Ibrahim Dawaye	1265	2442	1376	5994	96304	48618	144922
Mal. Kani Dankoli	1221	2797	1243	4662	90887	51282	142169
Alh. Abdul Azeez Usman	1243	2264	1399	5772	95726	45510	141236
Alh. Bala Atara	1265	2553	1088	4218	89377	46371	136108
Usman Nuhu	1088	2220	1399	6260	87956	45821	133777
Alh. Yusuf Mai Zare	1066	1709	1265	5994	83650	37629	121279
Rabilu Tsibiri	1043	1598	1265	5994	82540	35964	118504
Mal. Sale Aminu	1177	2398	Not	Planted	58830	35964	94794
Mal. Aminu Dugurawa	733	1154	244	2220	42491	21756	64247
Mean	1196	2623	1241	5408	85318	48608	133926
Control 1 (13 samples)	307	1109	732	1541	37418	19717	57135
Control 2 (4 samples)			1355	1712	32520	3424	35944
Based on 2003 prices	Grain	Fodder			Grain	Fodder	
Cowpea	N50/kg	N15/kg		Cereal	N24/kg	N2/kg	

therefore, all the participating farmers made a profit and were happy with the new system.

Some available results from the mass dissemination trials conducted during the 2002 to 2004 crop seasons indicate gross incomes ranging from Naira 102,000 to Naira 167,000 in the improved systems compared with Naira 24,000 to Naira 52,000 in the traditional systems (Figure 1, Table 6). The total cash inputs for improved seeds, fertilizers and chemicals in the improved systems ranged from Naira 8,000 to Naira 10,000. Thus, participating farmers greatly benefited from the improved system and they were extremely happy with the economic gains and family food security emanating from the project. Farmers are able to settle their debts, repair their houses, ensure better diets for the family, send their children to school and, after meeting those needs, many of them are also able to purchase land, motorcycles, water pumps, work

FIGURE 1. The improved 'sorghum-cowpea' and 'maize-double cowpea' strip cropping systems in farmers' fields.

A: Traditional 1 row sorghum : 1 row cowpea intercropping with little or no inputs.
B: Improved 2 rows sorghum : 4 rows cowpea strip-cropping with minimum and selective application of inputs.
C: Improved 2 rows maize : 4 rows double crop cowpea strip-cropping with selective inputs.

TABLE 6. Productivity of improved cowpea-based systems in northern Nigeria.

System	No farmers	Cowpea yield		Cerels yields		Total naira* Mean value
		Mean Grain	Kg/ha Fodder	Mean Grain	Kg/ha Fodder	
Kano State-2002						
Maize-cowpea (impd.)	39	731	2909	769	1731	102,103
Traditional (not available because maize is not grown in dry savanna)						
Kano State-2003						
Sorghum-cowpea (impd.)	349	1137	1942	688	2429	107,350
Traditional	29	301	834	786	1308	49,404
Kano State-2004						
Sorgh.-cowpea (impd.)	211	1025	2221	1333	5415	127,387
Millet-cowpe (impd.)	131	1008	2918	1132	4424	130,186
Traditional	78	135	389	264	2412	23,745
Kaduna State-2002						
Maize-cowpea (impd.)	55	1693	3894	893	1647	167,786
Traditional (not collected)						
Kaduna State-2003						
Maize-cowpea (impd.)	192	1757	1333	1333	1711	153,004
Traditional	43	458	993	993	1274	52,085
Kaduna State-2004						
Maize-cowpea (impd.)	324	1743	1202	1612	2139	160,588
Traditional (maize-cowpea)	9	541	733	544	1038	57,386
Traditional (sole maize)	12	-	-	833	976	24,034

Impd. = Improved system; *Prices (2004): Cowpea grain = N50/kg, cowpea fodder = N15/kg, cereals grain = N2/kg, cereals fodder = N2/kg, groundnut grain = N36/kg, groundnut fodder = N15/kg.

bulls, etc. The women participating farmers are also making enough money that they are now free from total dependence on their husbands.

DISCUSSION

Evaluation of different intercrop-planting patterns has shown that a 2 rows cereal-4 rows cowpea strip-cropping arrangement is much superior and it seems to hold great promise for increasing food production in West Africa without affecting the environment and degrading the soils. This is based on the introduction of new, more productive dual purpose and resilient cultivars of cowpea, maize, sorghum and millet in strip-cropping patterns involving 2 rows of cereals and 4 rows of cowpea with single or double crop cowpea, depending upon the rainfall. The superiority of the system emanates from several factors. The two cereal rows have no competing border rows and therefore, they yield the equivalent of almost three rows. Cowpea does not suffer competition from cereal rows because of its early maturity and slow initial growth of the cereals. Cowpea fixes atmospheric nitrogen and causes suicidal germination of *Striga hermonthica*. The increased grain production from this system ensures household food security as well as cash income and the crop residues are fed to livestock kept on the compound, which permits on-farm collection and use of manure to enhance soil fertility and organic matter. The two-third cowpea and one-third cereal combination minimizes fertilizer use and maximizes profit because of the higher prices of cowpea grain and fodder and at the same time, it leaves a positive residual soil nitrogen balance (Carsky et al., 2001) and reduces *Striga hermonthica* seed bank, both of which benefit the cereal crops. This combination involving less cereals and more cowpea is also appropriate in view of the global surplus of cereals and global deficit of legumes. For example, Nigeria imports more than 400,000 tons of cowpea each year. The one-third area under cereals ensures adequate production for home consumption and the internal Nigerian market and the extra cowpea production from the two-third area should reduce the cowpea import. The increased cowpea and cereal grain production ensures higher income and household food security and an increased production of cowpea haulms helps feed more livestock during the dry season, leading to a greater crop-livestock integration, which together have potential for reducing poverty and improving the quality of life of African farmers.

Some Success Stories

Hajia Hindatu Musa is one of the participating women farmers from the Kano State. She is the secretary of the registered Gamaryawa Women Farmers Association in Garko. She states, "I have been farming for the past 36 years, but what I realized from growing improved cowpea in the last three years surpasses all my 36 years of farming put together." Hajia is thankful to IITA for bringing this life-saving/improving technology to her village.

Alhaji Rabilu Sule is the leader of the Gwagwaranda Fadama Farmers Cooperative Society of Gwagwaranda village in Kano State. Through the extra income from the new system in two years, he purchased two work bulls, 16 goats, six sheep, eight poultry and one donkey–all of them generate several truckloads of manure. He states, "My family now lives a happy and healthy life, thanks to the new cropping system."

Adamu Abubakar rears small ruminants in his village of Kausanni in Kano State. He had always rented a work bull for his land preparation. After selling about 1 ton of cowpea seed produced using the new system in 2004, he was able to raise money to buy two work bulls. He comments, "I can now cultivate more land to produce enough food to feed my family and also make money by renting out the work bull."

Aminu Yusuf, a farmer from Wudil, Kano State, joined the project in 2004 as a result of the bumper harvest of cowpea produced by his neighbors who participated in the project. He produced about 1.2 tons of cowpeas from a rented piece of farmland. He sold part of the produce and bought his own farmland and some sheep.

Halidu Mohammed of Kausanni Cooperative Association resides along the Wudil riverbank in Kano State. He realized about 1.8 tons of improved cowpea in 2004. He sold some of it and bought a pumping machine and other facilities needed for dry-season vegetable production. He has high praise for IITA, Gatsby Foundation and USAID for extending the early-maturing and high-yielding cowpea varieties to his village.

Hajia Rabo Abdulrahman is the President of Albalka Women Farmers Association in Yakasai Area of Kano State. After selling her cowpea seed produced from the new system in 2004, she was able to settle some of her domestic needs and from the remainder of money, she purchased a piece of farmland. She expressed, "I never dreamt of ever having a field of my own. Now I have one, thanks to IITA, and Gatsby Foundation."

Malum Hassan NaGambo is one of the participating farmers in Giwa Local Government Council (LGC) of Kaduna State. In 2002, he planted an IITA-improved cowpea variety on a 0.4-hectare plot. He was so thrilled with the bumper harvest that he decided to increase the area to one-hectare in 2003. From the sale of his first cowpea harvest alone, he made about Naira 50,000, from which he paid his debt of Naira 7,820 and purchased a calf worth Naira 15,000. He obtained a good harvest from his second cowpea and maize crop from the same field and purchased another calf. He planted several hectares under the improved system in 2004 and 2005.

Alhaji Ahmed Galadima is one of the IITA contact farmers in this project. Through the extra income generated by use of the improved technology, Alhaji Galadima today owns nine goats, nine sheep, two cows and one donkey, as well as one milling machine, a wheelbarrow and a bicycle. He proudly says, "With this number of animals and the milling machine, I generate more income and grow more cowpea each year to generate enough food and fodder to feed them." Eight of his children now go to school, and his family is fully food-secure.

Rabi Abdulrahman, a woman farmer in Bichi, Kano states, "I depended on my husband for everything. And with the household having more than one wife, there was a strain on who was to get the provisions and who was to be left out. Now that I have income from the improved cowpea, I am able to make decisions concerning the household without bothering my husband. In the past my husband would buy standard books to be shared among all the wives' children, but I am now able to buy specialized books that can help my children. I am also able to supplement my children's diet with additional food that I grow myself."

Asamahu Mai, another woman farmer in Bichi, Kano state, expressed her sentiments as: "Before I used to rely on stitching caps for the men. This took very long and the returns were 60 Naira for a cap that I sewed in 3 days. Now with extra income from improved cowpeas, I am able to earn enough money that I now have a sewing machine and I can make additional items for sale. My children are eating well and showing less nutritional deficiency. With the improved cowpea I now have enough produce for the house and for the market."

Mallam Rawal Abdullahi is one of the farmers in Giwa Local Government Area (LGA) of Kaduna State. Narrating his story to Drs. Yvonne Pinto and Laurence Cockcroft from the GATSBY Foundation, he proudly showed his zinc-roofed house. His house is the only house roofed with zinc in a vast extended family compound. The house was roofed with proceeds from the sale of an improved cowpea. He also showed the goats that he purchased and the new livestock pen and the manure already generated through the feeding system that the project introduced to him.

Mallam Abdullahi Ibrahim, also of Layin Taki and a third-year farmer in the project, showed his two work bulls purchased from sales of cowpea produced under the project. He bought two goats in 2002 and now he has eight goats. He also bought one bull in 2003 and another one in 2004. According to him, he now generates sufficient manure at his house and his demand for inorganic fertilizer has gone down.

General Empowerment of Men and Women Farmers

The participating men and women farmers, through their registered groups, are already feeling more empowered and contacting the local authorities for greater services in the villages. For example: (1) the Women's Group in Garko LGA obtained community land for improved crop production by the group, and (2) the Yakassai Men's group obtained tractor service and animal-drawn ridgers at subsidized rates from the chairman of local government. More than 1000 farmers were given cash credit through their cooperatives from the NACRDB (National Agricultural Cooperative and Rural development bank) facilitated by the Wealth Windows foundation of Nigeria.

Family, Social and Educational Benefits

The emergence of women farmers' groups through this project has made history in the Hausa land where women are traditionally confined to houses. They are now able to come together, discuss various issues, including health and family, jointly plan social and economic activities and feel more empowered, respected and recognized. One of the women members, Maijida Balaraba Ado, now fearlessly sends her 14-year old daughter to secondary school even though her husband wanted to marry her off. These women freely attend field days and talk about their success stories in public.

REFERENCES

Baker, E.F.I. and D.W. Norman. 1975. Cropping systems in northern Nigeria, IAR, Samaru, Nigeria.

Carsky, R.J., B.B. Singh and B. Oyewole. 2001. Contribution of early season cowpea to late season maize in the savanna zone of West Africa. Biological Agriculture & Horticulture 18:303-316.

Edwards, R. 1993. Traditional farming systems and farming systems research, pp. 95-108. In: J.R.J. Rowland (eds.), Dryland Farming in Africa. The Macmillan Press, Ltd. Hong Kong.

Henriet, J., G.A. van Ek, S.F. Blade and B.B. Singh. 1997. Quantitative assessment of traditional cropping systems in the Sudan savanna of northern Nigeria. I. Rapid survey of prevalent cropping system. Samaru J. Agri. Res. 14:27-45.

Mortimore, M.J., B.B. Singh, F. Harris, and S.F. Blade. 1997. Cowpea in traditional cropping systems, pp. 99-113. In: B.B. Singh, Mohan Raj, K. Dashiell and L.E.N. Jackai (eds). Advances in Cowpea Research. Co-publication of International Institute of Tropical Agricultural Science (JIRCAS). IITA, Ibadan, Nigeria.

Norman, D.W. 1974. Rationalizing mixed cropping under indigenous conditions: The example of northern Nigeria. Journal of Development Studies 11:3-21.

Sanginga, N., K.E. Dashiell, J. Diels, B. Vanlauwe, O. Lyasse, R.J. Carsky, S. Tarawali, B. Asafo-Adjei, A. Menkir, S. Schulz, B.B. Singh, D. Chikoye, D.

Keatinge and R. Ortiz, 2003. Sustainable resource management coupled to resilient germplasm to provide new intensive cereal-grain-legume-livestock systems in the dry savanna. Agriculture Ecosystems and Environment 100: 305-314.

Singh, B.B. 1994. Breeding suitable cowpea varieties for west and central African savanna, pp. 77-87. In: J.M. Menyonga, Taye Bezuneh, J.Y. Yayock and I. Soumana (eds.). Progress in Food Grain Research and Production in Semi-Arid Africa. OAU/STRC-SAFGRAD Ouagadougou, Burking Faso.

Singh, B.B., O.L. Chambliss and B. Sharma. 1997. Recent advances in cowpea breeding, pp. 30-49. In: B.B. Singh, Mohan Raj, K. Dashiell and L.E.N. Jackai (eds.), Advances in cowpea research. Copublication of IITA and JIRCAS, Ibadan, Nigeria.

Singh, B.B. and H.A. Ajeigbe. 2002. Improving cowpea-cereals-based cropping systems in the dry savannas of West Africa, pp. 278-286. In: Fatokun, C.A., S.A. Tarawali, B.B. Singh, P.M. Kormawa, and M. Tamo (eds.), Challenges and opportunities for enhancing sustainable cowpea production. IITA, Ibadan, Nigeria.

Singh, B.B., P. Hartmann, C. Fatokun, M. Tamo, S.A. Tarawali and R. Ortiz. 2003. Recent progress in cowpea improvement. Chronica Horticulturae 43:8-12.

Terao, T., I. Watanabe, R. Matsunaga, S. Hakoyama and B.B. Singh. 1997. Agro-physiological constraints in intercrop cowpea: an analysis, pp. 129-140. In: B.B. Singh, Mohan Raj, K. Dashiell, and L.E.N. Jackai (eds.), Advances in cowpea research. Copublication of IITA and JIRCAS, Ibadan, Nigeria.

van Ek, G.A., J. Henriet, S.F. Blade and B.B. Singh. 1997. Quantitative assessment of traditional cropping systems in the Sudan savanna of northern Nigeria II. Management of Productivity of major cropping system. Samaru J. Agri. Res. 14:47-60.

doi:10.1300/J411v19n01_08

Bacterial Wilt and Drought Stresses in Banana Production and Their Impact on Economic Welfare in Uganda: Implications for Banana Research in East African Highlands

S. Abele

M. Pillay

SUMMARY. This study investigates the economic impact of banana *Xanthomonas* wilt (bxw) and drought on banana production in Uganda. The objective of this research is to determine the benefits of targeted research to avoid economic losses. In the worst-case scenarios, spread of bxw at a rate of 8% per annum, or drought at 50% yield losses in a five-year interval, results in significant losses for both consumers and producers. These losses would not only seriously jeopardize food security, but also affect overall macro-economic performance in Uganda. More likely scenarios with lower bxw and drought losses still show high economic losses, but they are mainly occurring on the consumers' side. Producers benefit from price increases at small production losses. This

S. Abele (E-mail: s.abele@cgiar.org) and M. Pillay (E-mail: m.pillay@cgiar.org) are affiliated with the International Institute of Tropical Agriculture, P.O. Box 7878, Kampala, Uganda.

[Haworth co-indexing entry note]: "Bacterial Wilt and Drought Stresses in Banana Production and Their Impact on Economic Welfare in Uganda: Implications for Banana Research in East African Highlands." Abele, S., and M. Pillay. Co-published simultaneously in *Journal of Crop Improvement* (Haworth Food & Agricultural Products Press, an imprint of The Haworth Press, Inc.) Vol. 19, No. 1/2 (#37/38), 2007, pp. 173-191; and: *Agricultural and Environmental Sustainability: Considerations for the Future* (ed: Manjit S. Kang) Haworth Food & Agricultural Products Press, an imprint of The Haworth Press, Inc., 2007, pp. 173-191. Single or multiple copies of this article are available for a fee from The Haworth Document Delivery Service [1-800-HAWORTH, 9:00 a.m. - 5:00 p.m. (EST). E-mail address: docdelivery@haworthpress.com].

doi:10.1300/J411v19n01_09

implies that research has to focus on public goods that can be delivered at no cost to farmers, as farmers under these circumstances are not likely to adopt costly preventive management measures. The best bet in this case would be publicly financed breeding, plant material multiplication and dissemination. Other options may be quarantine or trade restrictions, however, research on linkages between trade and the spread of bxw is yet to be done. doi:10.1300/J411v19n01_09 *[Article copies available for a fee from The Haworth Document Delivery Service: 1-800-HAWORTH. E-mail address: <docdelivery@haworthpress.com> Website: <http://www.HaworthPress.com> © 2007 by The Haworth Press, Inc. All rights reserved.]*

KEYWORDS. *Musa*, impact, banana markets, Uganda, *Xanthomonas* wilt, breeding

INTRODUCTION

The East African Highland banana (*Musa* spp. AAA), in particular the cooking type, is one of the most important staple food crops in Uganda and the adjacent countries, in particular Rwanda and Burundi. Ugandans consume 185 kg of cooking bananas per capita per year, comprising one-third of the caloric intake from starchy staples (FAO, 2004). A household spends about 35-50% of the household food budget on banana consumption (Kiiza et al., 2004). Bananas are also an important income source for about 30% of the Ugandan farmers as cooking bananas are marketed at a rate of 25-50% of production (Okech et al., 2004a). Uganda is one of the largest banana producers in the world, being second only to India. Trade in bananas is particularly important between Uganda and Rwanda, where 400,000 metric tons are traded annually (FAO, 2004).

Consequently, any threat to banana production in Uganda is likely to jeopardize livelihoods, food security and the overall economic performance of the Ugandan economy. Recently, a number of such threats have emerged. Besides abiotic problems, such as soil degradation and biotic constraints such as nematodes, weevils and fungal diseases, other emerging biotic and abiotic risks are serious threats for the sustained production of bananas in Uganda. One of them is banana *Xanthomonas* wilt (bxw) and the other one is drought. The two factors impose unique risks since there are no resistant cultivars (in the case of bxw), control measures are expensive, or the occurrence of the problem is erratic, in the case of drought/water stress. Consequently, research is needed to seek means to manage these stresses in a viable and cost-efficient way for farmers.

Public research requires economic analyses on different levels of decision-making. Firstly, public research investments in a specific domain have to be justified in terms of their expected impact on livelihoods of the target groups. In other words, decisions have to be made on (a) the principal research domain (e.g., bananas vs. cassava, or single crop research vs. systems), (b) specific aspects within the chosen domain, i.e., a choice of research on diseases, agronomy, or drought in the banana domain, and (c) different research topics, like crop management vs. breeding to combat a certain disease. On all levels, decision-making has also to be based on economic criteria, i.e., maximizing the potential economic welfare from interventions on the target groups.

The objective of this study is to evaluate the economic losses that might be induced by the effects of bxw and drought on the Ugandan banana sector. The concept is based on partial market analysis, with shifting supply and demand functions for cooking banana across time. It is assumed that yield losses due to bxw and drought induce supply reductions, which in turn induce increases in price and therefore losses in producer and consumer welfare, the former due to quantity losses, and the latter due to both price increases and quantity losses. The economic losses determine the investment potential in crop research, as the loss avoided represents the returns on investment for public research. An assessment of economic losses, and their distribution across time and stakeholders should also indicate the type of technology to be developed by research, and the mode of its dissemination.

Beyond the analysis, the study also presents a tool for ex-ante impact evaluation of crop research topics. The model used allows comparing different research problems with respect to their magnitude in economic terms, because it catalyzes agronomic and crop science knowledge and transforms it into comparable economic figures.

MATERIALS AND METHODS

Basic Concept

The method applied in this study is based on the concept of economic surplus models derived from partial market analyses (Alston et al., 1995). It is assumed that consumers and producers gain rents on markets, the so-called consumer and producer surpluses, and that these rents are appropriate to measure economic welfare, or wealth. Markets are constituted of supply and demand, and any change in supply and demand affects consumer and producer surpluses and consequently wealth. While changes in demand are mostly due to price, income and demography, as well as prices for substitute and complementary goods, supply changes are–besides price changes–mainly due to new production technologies that increase supply (Henze, 1994), or in the case of this study, adverse events like biotic and abiotic stresses, such as pests, diseases and drought.

Consumer surpluses are defined as the difference between the consumers' willingness to pay and the actual market price at any point along the demand curve from zero to the equilibrium quantity, whereas producers' surplus is defined as the differential of the supply price and the actual market price at any point of the supply curve between zero and the equilibrium quantity (Henze, 1994).

Figure 1 depicts the welfare effects of a supply shift induced by a stress scenario as applied in this study. Due to a stress factor, supply represented by the supply function S_1 is reduced to S_2. At a given demand function, the market equilibrium shifts from q_1,p_1 to q_2,p_2, reducing economic welfare. Producer surplus is reduced from the initial triangle ap_1b to dp_2e, whereas consumer surplus is reduced from acp_1 to dcp_2.

FIGURE 1. Welfare Impact of a Supply Shift

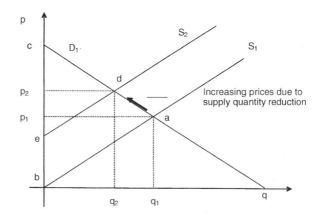

Adapted from Quaim (1999).

Calculation of Demand and Supply Shifts

It is clear from the Figure 1 that the shifters, both the supply and the demand shifter, have to be determined. It should be noted also that demand shifts across time, as stated above, due to population and income developments, as well as substitute price changes. The basic concept for computing the new equilibrium $q_2 p_2$ is described by the following formula (Henze, 1994):

$$dq_s{}^* = dq_s + e_{qp}dp \qquad (1)$$

with
$dq_s{}^*$ being the overall supply change
dq_s being the (positive or negative) supply change due to supply shifts
e_{qp} denoting the price elasticity of supply, and
dp denoting the equilibrium price change.

The demand shift effect is computed accordingly as:

$$dq_d{}^* = dq_d + \eta_{qp}dp \qquad (2)$$

with
$dq_d{}^*$ being the overall demand change
dq_d being the (positive or negative) demand change due to demand shifts
η_{qp} denoting the price elasticity of demand, and
dp denoting the change of the equilibrium price.

dp and the new equilibrium quantity and price (q_2 and p_2 in Figure 1) are obtained by equalizing the new demand and supply curves obtained from Equations 1 and 2.

Calculation of Welfare and Welfare Changes

Welfare is calculated as the consumer and producer surplus as depicted in Figure 1. In this concept, the consumer surplus is denoted by the integral of the demand function minus the turnover at equilibrium price and quantity, i.e., the sum of all demand price − equilibrium price differentials as stated above:

$$\int_0^{q*} D(q)dq - p*q* \tag{3}$$

With p* and q* being the equilibrium price and quantity.

The producers' surplus is computed accordingly as the turnover minus the integral of the supply function at equilibrium price and quantity:

$$-\int_0^{q*} S(q)dq + p*q* \tag{4}$$

Databases

Demand Function and Shifters

Demand shifts in the model are determined by exogenous factors (shifters) like income and demographic changes. The demand function in the model is based on a per capita demand function of banana kg price, per capita income and the square of per capita income, denoting decreasing demand for the staple at increasing income (negative income elasticity). The demand function was estimated on the basis of a time series from 1992 to 2001 (Table 1). The demand shift is determined by the changes in per capita income and population growth, the latter shifter computed by aggregating per capita demand according to population.

Supply Function and Shifters

The supply function has been estimated as a function of price and population, the latter variable denoting subsistence production and production increases due to increasing turnovers and profits even at declining prices. This supply function was estimated on the basis of production time series from 1992 to 2001 (Table 2). Technology-related supply shifters are more difficult to determine than demand shifters. In some studies, they are estimated as temporal functions, like logistic adoption functions (Quaim, 1999; Abele et al., 2005) or trend models (Kayobyo et al., 2005). In this study, the negative supply shifters are determined by the yield losses that are due to bxw and drought, as well as the exogenous population-induced supply shifter as a positive shifter. The basic parameters are shown in Table 3.

TABLE 1. Demand Function Parameters

Variable		Coefficient	T value (sig.)
Per capita consumption	DEP		
Price		−0.640000	−1.511 (0.169)
Income (per capita)		0.643000	3.341 (0.010)
Income squared		−0.000324	−2.850 (0.017)

R^2 adjusted 0.998 (Regression through origin). N = 10.

TABLE 2. Supply Function Parameters

Variable		Coefficient	T value (sig.)
Supply quantity	DEP		
Price		5261	2.649 (0.029)
Population		184	10.250 (0.000)

R^2 adjusted 0.999 (Regression through origin). N = 10.

TABLE 3. Supply and Demand Function Parameters

Parameter	Value
Demand function	
Price elasticity of demand at baserun	−0.600
Income elasticity at baserun	−0.001
Population growth factor	1.030
Real p.c. income growth factor	1.025
Supply function	
Price elasticity of supply at baserun	0.200
Population growth factor	1.030
Scenario shift factors	
Annual reduction rate of production induced by bxw	0.080
Reduction rate through drought	0.500

Adapted from FAO (2004), OECD (2004), Spilsbury et al. (2002).

Analysis

Baserun

The baserun is an extrapolation of market developments, i.e., prices and quantities during the next decade, by means of the market model. It is based on the "ceteris paribus" assumption, meaning that all other factors, except the ones depicted as growth factors (i.e., population and income growth factors) in Table 1, remain constant. The factors included in the analysis are the ones depicted above, i.e., supply and demand and their respective shifts without the scenario shift factors, i.e., possible shifts induced by production losses through disease and drought. The results of the baserun in terms of quantities, prices and consumer and producer surpluses serve as the basis for comparison of the scenarios.

The baserun observes declining prices of about 55% by the end of the decade, from 173 Ugandan shillings (UShs)/kg to 77 UShs/kg. Quantities produced and consumed increase from 4.5 million to slightly above 5.1 million tons representing a change of about 13%. The decline in prices is explained by the negative income elasticity and the low price elasticity of demand. The result of the baserun is depicted in Figure 2.

RESULTS

Scenario 1: Uncontrolled bxw

Bxw first occurred in central Uganda in 2001. It has now been reported in 34 districts, apparently spreading from the less intensive ba-

FIGURE 2. Banana Market Developments in the Baserun

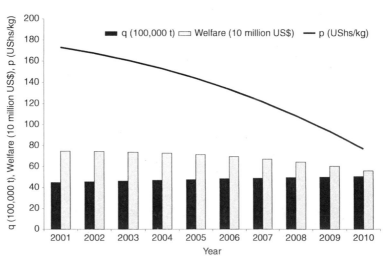

nana-production districts in central Uganda to the high-production areas in Western Uganda. Whereas in central Uganda and in the Ugandan districts adjacent to DRC in the West and to Kenya in the East infestation rates reach levels of 18-27%, the major banana producing areas in the South-West of Uganda show little or no infection (Tushemereirwe and Opolot, 2005). Kayobyo et al. (2005) reported that if uncontrolled, bxw spreads at an infection rate of 8% per annum. This adds up to a total production loss of bananas of about 56% during a 10-year period. However, despite the projected losses, both farmers' awareness and responses to bxw are not yet at a desired level, as adoption of preventive management techniques lags behind the expectations. The reasons for this remain unknown. It appears that in less affected areas farmers are less aware and uninterested in management techniques for bxw, which have a preventive character at zero or low infestation rates (Tushemereirwe and Opolot, 2005).

The effects on the markets are depicted in Figure 3. Prices increase significantly, especially during the first few years, with hikes of 240 UShs/kg. Exogenously given quantities decrease by 53%. Welfare is less than half (250 million US$) of the baserun at the end of the decade.

The welfare losses for both producers and consumers are depicted in

FIGURE 3. Banana Market Developments in the Uncontrolled bxw Scenario

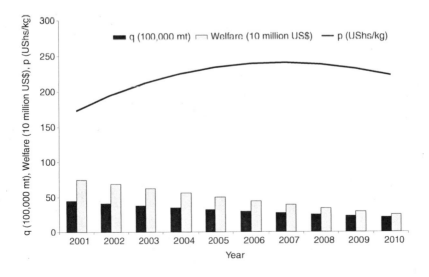

Figure 4. It is quite interesting to note that in the first two or three years, producers actually gain from a bxw epidemic, as they obtain higher prices and do not suffer much quantity losses. This may be a good reason why the uptake of bxw-management options among farmers is currently rather low: pest and disease management in general is mainly curative, i.e., it is applied only when the pest/disease occurs and a certain economic damage threshold is reached (Peterson and Hunt, 2003). The problem with bxw is that there is no curative management practice available, and that all practices are merely preventive. However, the incentives for preventive measures are low, if the farmers do not experience losses. Such a threshold is not yet reached, in fact, the disease does not occur yet, and profitability for farmers even increases, so that there are little or no incentives to adopt preventive measures. As time elapses, producers lose more, and, as their households are at the same time consumers, are also affected by consumers' losses. This may then result in a higher uptake of bxw-preventing management technologies.

The overall economic loss due to bxw equals 2 billion dollars across 10 years, which is an average of 200 million US$ per year. Although most of the losses occur on the consumers' side, producers are also affected by consumer losses since they are at the same time producers and consumers.

FIGURE 4. Welfare Effects of Uncontrolled bxw

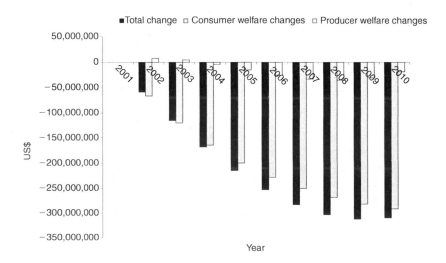

Scenario 2: Drought

Although drought was not considered a serious problem in the Lake Victoria Basin in the past, it has now become more frequent and is a matter of concern (Government of Uganda, 2002). Okech et al. (2004b) showed that yield reduction due to drought can reach 50% in some areas In our model, we assume that severe drought occurs twice within a decade, in the third and eight year, respectively, with a yield loss of 50%.

Figure 5 depicts how the market will respond under drought conditions. Prices, quantities and welfare changes basically follow the baseline scenario, as drought is erratic and has only an effect in the respective year. Price hikes are observed under drought conditions since the equilibrium quantity is reduced by 50%. The first price peak is at 250 UShs/kg, implying a 43% price increase, while the second peak is lower, due to the demand decrease across time and reaches only 199 UShs/kg, implying a 63% price increase.

Figure 6 shows the economic losses of a "drought only" scenario. It shows that the consumers are the major losers. Producers lose out only in the first drought period, whereas in the second drought period, they gain a little. The overall losses due to drought are smaller than those of a persistent bxw disease and amount to approximately $400 million across a decade, assuming two drought periods.

Scenario 3: Uncontrolled bxw-cum-Drought

The *bxw-cum-drought* scenario assumes a persistent bxw epidemic in conjunction with drought in the same intervals as above. Market responses are depicted in Figure 7. Basically, the markets behave similarly as in the bxw scenario, with two price peaks along the drought seasons. However, prices now reach 327 and 304 UShs/kg in the first and second drought season, respectively. Price volatility is, however, higher in the first drought season, with an increase of 68%, whereas in the second drought season, only a 26% increase is observed.

Economic losses in this scenario add up to about 2.6 billion US$ across a decade, or an annual gross domestic product (GDP) loss of 4%. It is interesting to note that this scenario's economic losses are higher than the sum of the above two scenarios, indicating that disease-affected agricultural systems are economically more vulnerable to short-term risks (which is actually expected). The scenario is depicted in Figure 8. It is also interesting to see that the producers' losses in a drought-cum-bxw scenario are much higher and that there are no

FIGURE 5. Banana Market Development in the Drought Scenario

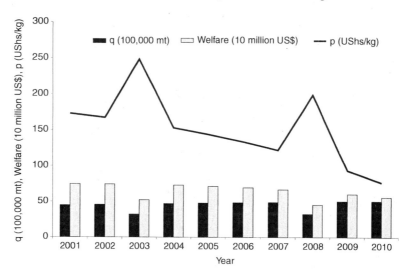

FIGURE 6. Welfare Effects of Drought

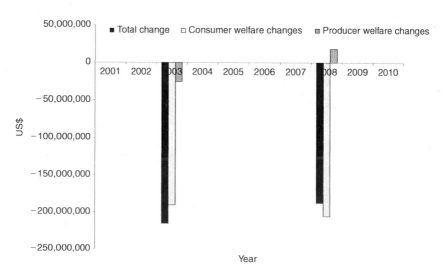

FIGURE 7. Banana Market Developments in the *Drought-cum-bxw* Scenario

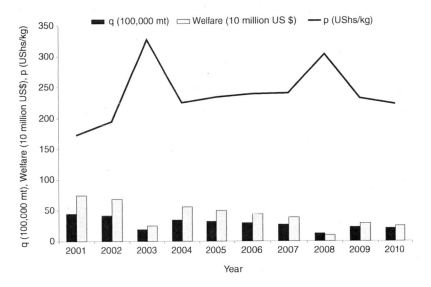

FIGURE 8. Welfare Losses in the *Drought-cum-bxw* Scenario

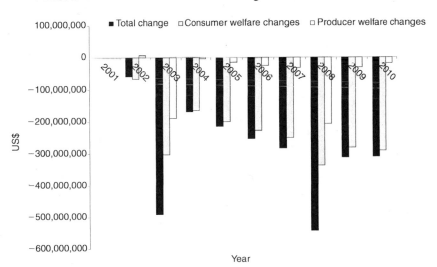

short-term producers' gains from drought-induced price increments, as in the drought only scenario.

Scenario 4: Different Intensities of the bxw Epidemic

The above are representative of 'worst-case scenarios' and provide a good overview of what may happen, especially in the case of an uncontrolled spread of bxw. They also point to a few insights that may be topics of further research. One interesting point is the obvious asymmetry between consumers' and producers' losses. The uncontrolled bxw scenario clearly revealed that producers gain in the first years and lose out only in the later stages of the epidemic, whereas consumers lose from the onset of the disease. Producers are also affected by much smaller losses than those of consumers. It would be interesting to ascertain the relationship between the intensity of the bxw epidemic and its concomitant economic losses and how these losses are distributed between producers and consumers. The following scenario discriminates and analyses five sub-scenarios, four of which are based on different bxw epidemic intensities, in particular annual bxw yield losses of 2, 4, 6 and 7%, respectively, whereas the fifth one tries to approach the status quo of the Ugandan banana sector, combining a relatively low bxw prevalence–2% yield losses per annum–with small drought losses that are estimated to be realistic for the second half of 2005 and early 2006.

Figure 9 shows the development of consumer surpluses in terms of changes with respect to the base run. In all scenarios, the consumers lose out.

Figure 10 shows the development of producer wealth under different bxw infestations. At lower infestation rates, producers gain from higher prices at only moderate yield losses. Only at higher infestation rates, from 6% per annum onwards, producers start losing. This indicates that at the present rather low infestation rates, there is little incentive for producers to adopt measures of bxw control.

An Assessment of the Present Situation in Uganda

The last analysis step is not based on a scenario-approach but tries to replicate the present situation in Uganda and assesses the welfare effects in the status quo. The model here is again a *bxw-cum-drought* model, as this is at present observed in Uganda, yet with a lower disease incidence and a lower drought incidence. The model has been calibrated

FIGURE 9. Consumer Wealth Changes Under Different Bacterial bxw Intensitites

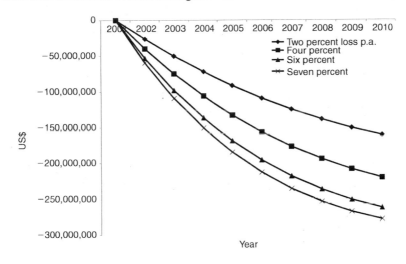

FIGURE 10. Producer Wealth Changes Under Different bxw Intensities

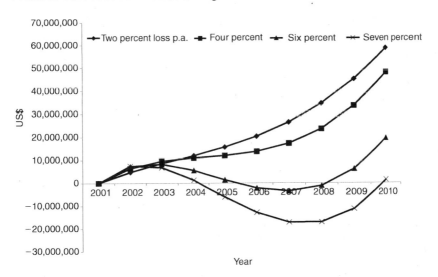

to replicate a price scenario that reflects the present market situation (FOODNET/MIS, 2006) by adjusting supply quantities in an observed framework of bxw prevalence and drought. The results are shown in Table 4 for the year 2006. The prospected average annual price is at 200 UShs/kg, which is about 50% higher than in the baserun, due to bxw prevalence and droughts in late 2005 and early 2006. Consumer wealth losses in comparison to the baseline scenario are at around 160 million dollars, whereas producers gain about 13 million dollars.

DISCUSSION

One of the major findings from the different scenarios is that *bwx* presents a potential threat of failure of Uganda's economy especially if the disease spreads throughout the major banana-growing areas and is not controlled. Left uncontrolled, bxw could potentially affect food security, small-scale farmers' livelihoods, and the macro-economic performance of Uganda. It is interesting to note that consumers suffer greater wealth losses than producers. Consumer losses imply a disparity not only between income groups but also between regions, since the market-dependent consumers are mainly urban dwellers. This presents a serious economic problem because the (relatively) high-income population is considered to drive the economy through demand for high value goods. The economic loss through bxw could trigger a negative spiral. Declining incomes could also induce a greater demand for staples like bananas, causing higher prices and greater welfare losses. Consequently, investments in bxw research and management will not only benefit the farmers and consumers by providing more income and food, respectively, but also the medium- and long-term macro-economic performance of Uganda. Drought as compared with bxw is certainly a lesser problem for the economy. Although drought causes a high absolute value loss in a worst-case scenario, this loss only accounts for 0.6%

TABLE 4. Market and Welfare Effects on Status Quo in Uganda (2006)

Item	Baserun	2006 status quo	Difference in %
Market price (UShs/kg)	133	200	50
Equilibrium quantity (t)	4,800,000	3,600,000	25
Consumer welfare to baserun	0	−157,418,661	n.a.
Producer welfare to baserun	0	13,486,622	n.a.

of the Ugandan GDP in a drought year. Problems due to drought can be mitigated by reducing problems caused by bxw since there is a positive relationship between drought and bxw.

The asymmetry between consumers' and producers' economic welfare (consumers losing more than producers who even benefit from higher prices) could have greater significance at more moderate–and actually more realistic–bxw and drought prevalence rates than shown in the above in the worst-case scenario. At lower prevalence rates, producers actually gain from higher prices at only small quantity reductions, whereas consumers still lose out. Consequently, the producers' threshold to prevent bxw is reached considerably after the consumers' and therefore overall economic threshold is reached. This means that the likelihood of farmers adopting especially preventive bxw management technologies is quite low unless a certain disease infestation threshold is reached, since they perceive their present situation as being better than before and consequently are not likely to invest in costly management-based control measures. The farmers' reaction threshold could be reached at an infection rate of 7% per annum, which is very close to the hypothesized worst-case scenario. However, with the anticipated consumers' losses and economic downturn, there is need for public research to combat bxw even before its main target group, the farmers, will have an economic incentive to react. But as the farmers are unlikely to react and adopt in a win-situation, research measures have to be carefully selected according to their potential of acceptance by the farmers. The basic criterion for research outputs is that their application can be publicly financed, i.e., no private investments in the case of adoption have to be made by the farmers, as they would not adopt in this case. On-farm management like debudding, or many other preventive IPM-strategies may not be an option, as they require inputs from farmers. A better option is breeding for resistant or tolerant varieties, including multiplication and dissemination of planting material. These activities can be publicly financed and introduced into farmers' fields whenever re-planting of over-aged mats is required. This would control the spread of bxw (and the negative effects of drought by providing more drought-resistant planting material) at no additional costs to the farmers and is therefore likely to be highly adopted. It is also a measure that does not affect markets, as every farmer has his own choice of replanting his optimal number of cultivars. Other public options to contain the disease may be trade restrictions and quarantine.

REFERENCES

Abele, S., Ntawuruhanga, P., Odendo, M., Obiero, H., Twine, E., and J. Odenya (2005): Effectiveness of breeding and disseminating CMD-resistant cassava varieties in Western Kenya, in: Proceedings of the 7th African Crop Science Society Conference in Entebbe, Uganda, from 7th to 9th December 2005, Kampala (ACSS), pp. 233-238.

Abele, S., van Asten, P., Gold, C., Pillay, M., and T. Dubois (2004): Applying economics on banana systems in ESA to target agronomic research. IITA work planning week 2004, book of abstracts. Ibadan: IITA.

Alston, J.M., G.W. Norton, and P.G. Pardey (1995): Science under scarcity; Principles and practice for agricultural research evaluation and priority setting. Ithaca and London: Cornell University Press, Ithaca and London.

FAO (2004): Statistical databases. www.FAO.org

FOODNET/MIS (2006): Market price information for Uganda.

Government of Uganda (2002): Initial National Communication of Uganda to the Conference of the Parties to the United Nations Framework Convention on Climate Change.

Henze, A. (1994): Marktforschung: Grundlage für Marketing und Marktpolitik. Stuttgart: Ulmer.

Kayobyo, G., Aliguma, L., Omiat, G., Mugisha, J., and S. Benin (2005): Impact of bxw on household livelihoods in Uganda. Paper presented at the workshop "Assessing the impact of the banana bacterial wilt *Xanthomonas campestris pv. musacearum* on household livelihoods in East Africa," held on Dec. 20th, 2005 in Kampala, Uganda.

Kiiza, B., Abele, S., and R. Kalyebara (2004): Market opportunities for Ugandan banana products: National, regional and global perspectives. *Uganda Journal of Agricultural Sciences*, 9(1): 743-749.

OECD 2004: Statistical databases. www.oecd.org

Okech, H.O., Gold, C.S., Abele, S., Nankinga, C.M., Wetala, P.M, van Asten, P., Nambuye, A., and P. Ragama (2004a): Agronomic, pests and economic factors influencing sustainability of banana-coffee systems of Western Uganda and potentials for improvement. *Uganda Journal of Agricultural Sciences*, 9(1): 432-444.

Okech, S.H., van Asten P.J.A., Gold, C.S., and H. Ssali (2004b): Effects of potassium deficiency, drought, and weevils on banana yield and performance in Mbarara, Uganda. *Uganda Journal of Agricultural Sciences*, 9(1): 511-519.

Peterson, R.K.D. and T.E. Hunt (2003): The probabilistic economic injury level: Incorporating economic uncertainty into pest-management decision making. *Journal of Economic Entomology*, 96(3): 536-542.

Quaim, M. (1999): Assessing the impact of banana biotechnology in Kenya. ISAAA briefs no. 10, ISAAA, Ithaca, NY.

Spilsbury D.S., J.N. Jagwe, and R.S.B. Ferris (2002): Evaluating the market opportunities for banana and its products in the principal growing countries of ASARECA (Uganda, Kenya, and Rwanda) country reports. Kampala: IITA/Foodnet.

Tushemereirwe, W. and O. Opolot (2005): bxw history, status and national strategies. Paper presented at the workshop "Assessing the impact of the banana bacterial wilt *Xanthomonas campestris pv. musacearum* on household livelihoods in East Africa," held on Dec. 20th, 2005 in Kampala, Uganda.

doi:10.1300/J411v19n01_09

Greening of Agriculture:
Is It All a Greenwash
of the Globalized Economy?

Charles Francis
Roger Elmore
John Ikerd
Mike Duffy

SUMMARY. Consolidation of farms, agricultural input supply companies, and commodity businesses over the past several decades have led to a concentration of ownership and control. There is growing concern by society about the environmental impacts of agriculture and the food system, and companies are eager to exploit this concern by advertising products that are environmentally friendly. When there is a *greening* of a company to reduce pollution or improve efficiency of non-renewable resource use, this is a legitimate way to justify advertising green products. When a company attempts to present a responsible public image, but

Charles Francis is affiliated with the University of Nebraska, Department of Agronmy & Horticulture, 279 Plant Science, Lincoln, NE 68583 USA and the Department of Plant & Environmental Sciences, Norwegian University Life Sciences, N-1432, Ås, Norway (E-mail: cfrancis@unlnotes.unl.edu).

Roger Elmore is affiliated with Iowa State University, 2104 Agronomy, Ames, IA 50011 USA.

John Ikerd is affiliated with the University of Missouri, Columbia, MO 65201 USA.

Mike Duffy is affiliated with Iowa State University, 478 Heady Hall, Ames, IA 50011 USA.

[Haworth co-indexing entry note]: "Greening of Agriculture: Is It All a Greenwash of the Globalized Economy?" Francis, Charles et al. Co-published simultaneously in *Journal of Crop Improvement* (Haworth Food & Agricultural Products Press, an imprint of The Haworth Press, Inc.) Vol. 19, No. 1/2 (#37/38), 2007, pp. 193-220; and: *Agricultural and Environmental Sustainability: Considerations for the Future* (ed: Manjit S. Kang) Haworth Food & Agricultural Products Press, an imprint of The Haworth Press, Inc., 2007, pp. 193-220. Single or multiple copies of this article are available for a fee from The Haworth Document Delivery Service [1-800-HAWORTH, 9:00 a.m. - 5:00 p.m. (EST). E-mail address: docdelivery@haworthpress.com].

does not change production practices we could call this a *greenwashing* or use of disinformation to mislead consumers. It is difficult to distinguish between the two.

This chapter explores two questions: is there a relationship between scale of farming and business and green activities, and does adoption of a multiple bottom line influence greening of agriculture and food systems? After examination of the effects of farm size, we conclude that consolidation may lead to less timely field practices, separation of day-to-day management and ownership, and reduced accountability to the local community. Yet there is conflicting evidence of whether smaller farms or businesses are greener than large ones. We are convinced that a farm and a business that measure success in terms of environmental soundness and social responsibility as well as economic returns will be greener than ones that only use economics as the single bottom line. This is consistent with our discussions with farmers through Extension meetings and other contacts as well as observations we make while visiting farms in the U.S. Midwest region. Everyone agrees on the need for a greener future, but there are differences among decision makers in agriculture and food systems about how to achieve this goal. doi:10.1300/J411v19n01_10 *[Article copies available for a fee from The Haworth Document Delivery Service: 1-800-HAWORTH. E-mail address: <docdelivery@haworthpress.com> Website: <http://www.HaworthPress.com> © 2007 by The Haworth Press, Inc. All rights reserved.]*

KEYWORDS. Sustainable agriculture, industrial agriculture, environmental impact, farm size, business size, monopolies, farm consolidation

---

Greening is the modification of a process or an organization to become more environmentally friendly, to reduce pollution, to improve renewable and non-renewable resource-use efficiency, and to conduct an activity in a sustainable manner (the authors).

Greenwashing is disinformation disseminated by an organization so as to present an environmentally responsible public image (Concise Oxford English Dictionary, 10th Edition).

INTRODUCTION

In today's complex agriculture and food systems, it is difficult to meet all the regulations and keep a farm or a food business profitable, as well as contribute to food security. In a global society that is facing limits on fossil fuels, ever-increasing population, and climate change, the issues of environment and long-term food security are high priority. Small steps in *greening* our food production and processing system will help, but it is essential to examine what is happening in the larger picture. We need to evaluate claims of *greening* by all players in the food industry, large and small, in order to decide if changes are making a difference, or if they are only *greenwashing* to enhance an image and increase sales.

There is fashionable rhetoric today about the desirability of family farms, local businesses and food systems, and how these are inherently *greener* than their industrial and global alternatives. Those in the sustainable agriculture community claim that large industrial-model farms, focused only on short-term profits, are obviously less environmentally sensitive than their small-scale counterparts. One argument is that organizations that evolved to maximize profits are inherently handicapped in attempts to improve the environment because of corporate goals and ownership structure (Waage and Francis, 2004). Those who support the industrial model, however, claim that larger operations can afford stewardship while feeding a growing population and providing cheap food. They can build on comparative advantage, a growing free market, and uninterrupted global trade. How do professionals sort out these claims and determine the effects of size and ownership on sustainability? Equally important, to what extent does the public understand and accept the impact that ownership and management strategies have on the *green ness* of agriculture and the food system?

The general public appears confused and even irritated about environmental issues, and advertising agitates both reactions. A recent "advertorial" from a major agricultural corporation promoting a herbicide claimed, "The goal is to kill all the weeds, because we know that dead weeds will not become resistant" [to the herbicide]. To the non-biologist, this may appear to be a reasonable statement. To the knowledgeable biologist, the suggestion is a *greenwash*. According to Bob Hartzler, Iowa State University weed specialist, "HerbicideX-resistant weeds will not die when exposed to HerbicideX; this is the definition of resistance." For example, annual application of a single herbicide, such as Roundup™, to the same crop or to a rotation of maize (*Zea mays* L.) and soybeans (*Glycine max* (L.) Merr.), when both are Roundup Ready™ (resistant to the herbicide), will lead to weeds that are resistant to this specific herbicide. The literature now reports 304 biotypes of 182 different species that have demonstrated resistance to one or more herbicides (http://www.weedscience.org/in.asp; 18 November 2005). The public can be confused by conflicting reports from experts.

Industrial scientists and farmers contribute to the confusion. Pesticide industry advertising describes how weed resistance is a minimal problem, and that rational use of recommended rates will solve the problem. Many farmers accept the argument, convinced that improved, more effective chemicals will replace current technology. Thus they can continue to gain maximum efficiency of labor use and lowest possible production costs. One result of this strategy will be larger farms. University scientists, however, often caution against the widespread use of any single chemical or technology that reduces biodiversity in the farming system, until the wider implications of each technology are more fully understood. The southern corn leaf blight in the early 1970s provides one example of the danger of wide deployment of a narrow germplasm base. Integrated systems education, for example agroecology, instructs us to look beyond simple issues, such as killing weeds, and consider production, economic, environmental, and social issues and impacts of the combination of broad technologies in the food system, including who benefits (Francis, 2004; Lieblein et al., 2005). Oblivious to the continuing debates in agriculture and academia, most people in society are unconcerned about where and how food is produced, as long as it remains inexpensive and available. We need to examine whether these aspects of the food system are important.

The retail food sector also sends promising signals about *greening*. From a recent announcement:

Newman's Own™ Organics at McDonald's™!

McDonald's™ now is selling Fair Trade Certified™ Newman's Own™ Organic Blend at 658 of its restaurants in New England. The company claims this is *great* news for farmers, for the environment, and for people who want high quality coffee. The advertising goes on the claim that the benefits of this partnership are endless, since the customer is helping many causes, including the farmer getting paid a fair price to grow a sustainable crop.

[Adapted from: www.greenmountaincoffee.com, accessed 4 November 2005]

On the surface, the credibility of the Newman's Own™ name and Fair Trade Certified™ label convey a *green* image for the product. In addition, the advice of The Natural Step™ and pressure from consumers have motivated McDonald's™ to replace styrofoam with paper packaging, and more recently, replace chickens raised by conventional methods with those raised antibiotic free. Perhaps it was coincidence, but these actions were taken shortly after the publication of the best-selling indictment of U.S. eating habits, *Fast Food Nation: The Dark Side of the All-American Meal* (Schlosser, 2001). It appears that the largest retailer of drive-through food is making a significant change. This is confirmed by a quick visit through the Yahoo!™ search engine to reveal the first three entries for McDonald's as (1) "McDonald's near you," (2) "McDonald's Hurricane Disaster Relief Update," and (3) "McDonald's Food and Nutrition easy-to-read tables of nutritional facts for popular menu items." The message is (1) convenience, (2) social responsibility, and (3) good nutrition [from Yahoo! search on 17 November 2005]. What the customer must decide is whether these moves are to gain market share and profits, a form of *greenwashing*, or to improve the environment? Or are they both?

Our purpose is to distinguish between *greening* and *greenwashing* and attempt to illuminate the hazy territory between them. We separate scientific fact from advertising fiction through reason, illustrated with examples and documented with published results. We address the influence of size and ownership patterns in agriculture and the food system, and we raise two fundamental questions:

- Is there a cause-effect relationship between scale of farming or business and *green-ness* of activities?
- Does the adoption of a broader set of goals in farming or business impact *greening* in agriculture and food systems?

We audit the triple bottom line that is being pursued by forward-looking farmers and companies: economic success, environmental accountability, and social impact, including distribution of benefits. The same factors have been used for the past two decades to describe sustainable agriculture (Francis et al., 1990). Through the discussion, we establish criteria for the public and for professionals to distinguish *greening* of agriculture from a *greenwash* of globalization.

Definitions of Greening *versus* Greenwashing, Industrial *versus* Sustainable *Agriculture*

There are many opinions about what is meant by these terms. In various contexts, they can create better insight for communicating ideas, or they can incite arguments if they are against individual or commercial interests. We call *greening* the modification of a process or an organization to become more environmentally friendly, to reduce pollution, to improve renewable and non-renewable resource-use efficiency, and to conduct an activity in a sustainable manner. *Greening* contributes to a secure food supply. In contrast, a *greenwash* is disinformation used by an organization to present an environmentally responsible public image, without the actual *greening* to back that up. In addition to the Oxford Dictionary definition, a similar one may be found, together with important links, on the Wikipedia web site: http://en.wikipedia.org/wiki/Greenwashing (30 January 2006).

It is also essential to define *industrial agriculture* and *sustainable agriculture*. For the purpose of this paper, we define *industrial agriculture* as having the goals of increased profits and concentrated ownership, including both consolidation of farms and mergers in the input and marketing sectors. The explicit goal is to maximize profits, usually in the short term. This is achieved by maximizing productivity through growth to reach higher sales, most often by increasing scale. There is specialization, standardization, and consolidation of control with limited internal constraints and controls. This strategy leads to larger farms and businesses that use both market and political power to remove external constraints to growth–including federal commodity subsidy programs, import tariffs, export subsidies, and other market-distorting measures that stimulate growth beyond actual physical economies of scale (Ikerd, 2005).

Definition of Industrial Agriculture

Industrial agriculture has the goal of increased and concentrated ownership, including both consolidation of farms and mergers in the input and marketing sectors. The explicit goal is to maximize profits, usually in the short term. This is achieved by maximizing productivity through growth to reach higher sales, most often by increasing scale (the authors).

It is useful to read the published statement of purpose of one large representative corporation in agriculture. For example, a statement summarized from the Monsanto web site says that Monsanto Company contributes through its products to improve environmental sustainability. These impacts include: *environmental impacts* ranging from eco-efficiency of manufacturing to yield optimization, resource conservation, and soil fertility; *economic impacts* such as biotech crops to improve profitability and well-being of large-scale and smallholder farmers; *societal impacts* creating safe and healthy work environments and community involvement (adapted from Monsanto, http://www.monsanto.com/, accessed 30 January 2006). The apparent triple bottom line is impressive, and later we examine the accountability of corporations to their mission statements when in practice their goal is to maximize profits.

In contrast, what we generally define as *sustainable agriculture* has primary goals of permanent productivity and minimal impact on the environment; its secondary goal is to contribute to the food supply at a local level. The strategies we observe sustainable farmers using lead to balance and harmony among three factors: economics, ecology, and social responsibility. A sustainable strategy on farm or in business maintains diversity, individuality, decentralization, and local control of food security within a framework of "sustainable capitalism" (Ikerd, 2005). Definitions are not unlike those of Monsanto, as we discuss later. The optimistic definition of Wendell Berry describes a long-term goal that would achieve an agricultural system that accepts disturbance of the ecosystem but is designed to maintain or build soils rather than deplete them.

Definitions of Sustainable Agriculture

"Sustainable agriculture addresses the ecological, economic, and social aspects of agriculture"

"An agriculture that does not deplete soils or people" [Wendell Berry]

[From http://www.leopold.iastate.edu/about/sustainableag.htm, accessed 4 November 2005]

More important than how terms are defined is how they are put into practice. To illustrate the application of these definitions, we provide examples, including those from an industrial corporation perspective. Many large companies today have corporate environmental policies, articulated in statements of intentions and principles in relation to overall environmental performance. As illustrated above, these statements provide a framework for a company's environmental objectives and targets. Since they are not required by law to follow these principles, and there is normally no transparency in the process of implementation, companies need not always comply with their own policies. Policies are not legally binding, and without implementation they can become a form of *greenwashing* (Ramos and Montiel, 2005). An exception is the Chicago Climate Exchange (CCX), an organization that attempts to incorporate legal accountability. From the description on their Web site, it appears that this group is implementing policy and establishing an economic basis for trading in greenhouse gases. Among the members of the Exchange are major agricultural input suppliers and industrial agricultural producers, including DuPont and Premium Standard Farms. CCX is the world's first and North America's only voluntary, legally binding rules-based system for greenhouse gas trading with a goal of reducing emission reduction [author's interpretation from their Web site, www.chicagoclimatex.com/, accessed 17 November 2005].

What drives a business corporation to environmental compliance with their policies when there is no economic advantage to do so? Ramos and Montiel (2005) focused on the oil and gas industry, chemical and non-chemical manufacturing, and the service sector, but the principles they present would appear to operate in most profit-oriented businesses. They suggest three motivators for complying with policy, not all of which operate in the contemporary business environment:

- *Coercion*: farm corporations or companies are forced to comply by regulations, laws, sanctions, or fines (examples are maintaining residue to qualify for federal farm payments, and limiting water or air pollution in emissions from factories).
- *Conformation*: corporations are morally governed by conformity to standards set by other companies or by the public (for example McDonald's move away from styrofoam containers because other fast-food companies had already taken this step, or because of public preference or pressure).
- *Imitation*: companies may want to look at least as good as the competitors (advertising ethics in the past included comparing one's product to "Brand X," but today there is frequent comparison of cost, function, or desirability, e.g., Coca-Cola™ versus Pepsi-Cola™).

Ramos and Montiel (2005) ask if there are any incentives to drive corporate environmental compliance with their policies when there is no economic advantage. They conclude, "economic advantage is most probably the motivator for companies to *implement* specific environmental policies. Our findings indicate that an outside stakeholder should look with a sceptical eye at any company that commits to a policy if that company does not have an economic motivation to do so." Their research further indicates that companies have embraced sustainability because they believe it could enhance business performance by reducing operating costs, developing innovative products and services (and a new image), and reducing the company's liabilities. Within this context, it is important to carefully evaluate individual or corporate statements of policy and ethics. To make rational choices in the marketplace, consumers need to observe what people or companies do, and not what they say.

MAJOR TRENDS IN AGRICULTURE AND FOOD SYSTEMS

Agriculture in the industrial world has developed toward increasing concentration of ownership in both crop and animal production, with strong trends toward fewer and larger farms, and similar industrial concentration in the input and grain trade corporations. Industrial development has the purpose of maximizing productivity by maximizing profits and growth. Profit is defined as total value of output minus total costs, maximized at a given point in time, and growth makes possible increasing productivity by increasing scale of operation, consistent with our definition above.

One example of contemporary change is illustrated by the current dilemma related to rapid growth in the organic food sector. The industrialization of growing and marketing organic food and control of this sector of the food system are moving out of the hands of the people whose social and philosophical roots were instrumental in starting organic farming (Ikerd, 2006). The system was generally oriented toward individual production and local food systems, but it is quickly becoming one small part of the global, industrial-model food system. Some claim that this has been accelerated by the standardization of organic certification rules, for example in the USA by the Federal Organic Act implemented in November 2002. This represents a departure from the loose affiliation of private and public certifying organizations coordinated by the global organization IFOAM (International Federation of Organic Agriculture Movements, http://www.ifoam.org/). Organic farming and food systems need to be profitable, at any level of scale, or they will not continue. What is being lost, according to founders of organic farming, are the social motivations about improving people's well-being.

In contrast to the industrial model, sustainable development is more than just economics. Although specialization, standardization, and hierarchal control may also be used to some degree on a sustainable farm, for example, the process is constrained internally by multiple goals as defined above (Ikerd, 2005). Those who started organic farming saw themselves as improving the environment, producing safe and healthy food, and generally filling an important role in society. Yet, the social responsibility issue has been questioned by Patricia Allen, who argues in *Together at the Table* (Allen, 2004) that there has been little attention toward treatment of labor, limited availability of organic food to a broader public, and a lapse in concern about equity of benefits. So there is question about whether the traditional organic agriculture community was achieving its stated goals.

Although some would maintain that it is self-evident that a sustainable development path in agriculture must by definition be *greener* than an industrial path, in this discussion we present an evaluation based on the literature and observation. We further suggest that only an informed consumer public can identify the characteristics of *greenwashing* or disinformation, and thus bring the power of the marketplace to establish a long-term, secure, and sustainable food supply. Current trends are relevant to the discussion, and we explore farm size and ownership, consolidation of seed companies, and market share of companies in the grain and livestock commodity business.

Farm Size and Ownership

During the last century, farm numbers in the USA have declined from about 6.4 million to just over 2 million farms today (USDA, 2002). This was a consequence of mechanization that "forced" a large part of the labor force to work in industry and other sectors of the economy. In the latter half of the 20th Century, consolidation was due to increased specialization in crops or livestock, to the availability of herbicides to reduce labor needed for weed control, and to chemical fertilizers that provided a way to maintain soil fertility and allowed separation of livestock from crop farms. In the Midwest, decline in farm numbers in Iowa (2.2 million in 1910 to 90.000 in 2004) and in Nebraska (1.3 million to under 50,000 in the same time period) was similar to the national trend (USDA, 2004).

From 1979 to 2003, average farm size increased only slightly in the USA and in Iowa, but much more in Nebraska (USDA, 2004). Average farm size is much larger in Nebraska (950 acres) than in Iowa (350 acres) due to lower rainfall and more extensive farming practices. The average numbers hide a strong trend in the loss of family farms that are near the average size. In the Midwest, as across the country, there has been a large growth in number of small hobby and specialty farms and a small growth in number of large corporate, industrial-model farms, and on average the farm size does not appear to change that much. We know from experience on the ground in Extension that there are fewer farmers coming to meetings on crop production practices, an indication of the loss of our traditional clients who manage family farms.

When we examine these changes in farm size and ownership, the data provide additional insight into rural change. Data from Iowa over the past two decades (Duffy and Smith, 2004) show that fewer existing farmers are purchasing farmland (Figure 1), in part because there are actually fewer farmers still in business who can afford to expand. There is little change in the numbers of new farmers purchasing land, varying between one and four percent, while the major growth is in purchase by investors. Perhaps even more important is that increasingly the investors reside outside the state of Iowa. Over the 20 years from 1982 to 2002, the proportion of land purchased by people residing outside the state climbed from 6% to almost 20%. One consequence is that ownership on more farms is separated from management and control (Duffy and Smith, 2004). Another change is the shift from sole owner and husband/wife ownership pattern toward other co-owners and trusts, perhaps a consequence of an aging farm population (Figure 2, adapted from Duffy and Smith, 2004). We do know that increased federal farm support programs for commodities have rewarded production, disproportionately favored larger farms, and driven up the cost of farmland and cash rents. Alternatives to this scenario have been proposed for decades, but so far to no avail (Cochrane, 2003; Levin, 2003).

Business Size and Ownership

Reported trends in farm numbers, farm size, and ownership patterns in the Midwest are similar to parallel changes in the farm input supply and the agricultural commodity sectors in the U.S. and worldwide. Both are relevant to our discussion of the *greening* of farming and industry related to agriculture. Nearly three decades ago, Dan Morgan's book *Merchants of Grain* (Morgan, 1979) revealed that only five corporations controlled over 80% of the international grain commodity trade.

FIGURE 1. Purchase of Farmland in Iowa by Current Farmers, by Investors, by New Farmers, and by Others from 1989 to 2004 (Adapted from Duffy and Smith, 2004).

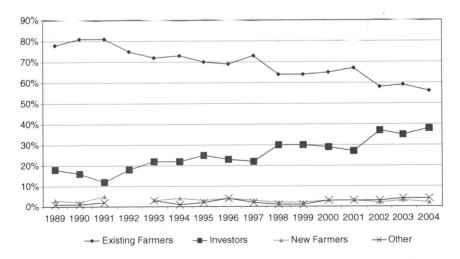

FIGURE 2. Changes in Farm Ownership Patterns in Iowa, 1982, 1992, and 2002 (Adapted from Duffy and Smith, 2004).

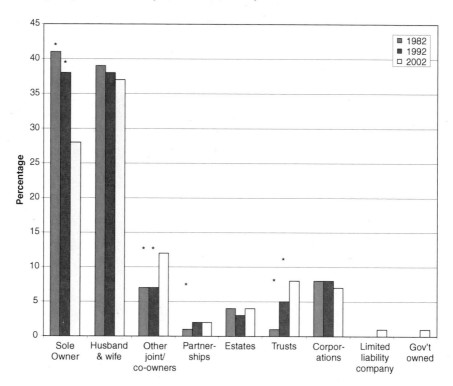

Heffernan and Hendrickson (2002) have been quantifying this trend for more than a decade. They calculate the concentration ratios in food processing as an example of growth in corporate market share. Latest calculations show that four corporations control 61% of grain handling, three firms control 81% of corn exports and 81% of soybean exports, four corporations control 61% of flour milling, and four corporations control 80% of soybean crushing. In the beef markets, four corporate packers control 81% of the market, four pork packers process 59% of all hogs, and four corporate broiler processors slaughter 50% of all chickens. The most dramatic increases in corporate concentration in recent years have been in food retailing where market control by the five largest corporations rose from 24% to 54% between 1997 and 2003. Heffernan and Hendrickson (2005) also describe the contemporary shift in power from governments to corporations. While national governments have a need to maintain political stability and to feed all of their country's people, corporations' chief mission is to maximize profits and thus increase wealth of their stockholders. There will be attention to environment and long-term sustainability only if it contributes to short-term profits.

One specific example relevant to the Midwest is the decline in number of U.S. seed corn companies from over 500 in 1940 to less than 100 in 2000 (Troyer, 2004). An indicator of this consolidation is the number of companies entering hybrids into the uniform performance tests conducted each year by Iowa State University across multiple locations. Over the past decade, the number of companies that submitted entries declined from 85 to 50 (J. Rouse, Iowa State University, personal communication, October 2005). This pattern of consolidation in seed corn companies is similar across the country.

Concentration of ownership in agriculture and related industries is part of a general trend in the USA that began in earnest in the 1970s according to labor journalist Sam Pizzigati (2004). He describes the inequalities in salaries that have arisen over the past three decades as a result of exemptions that reduce taxes for corporations under current law and provide no control of maximum wages. Industrial agriculture and corporations reflect this trend. In part to counter the negative publicity attached to high CEO salaries, some corporations claim to be giving attention to a multiple bottom line, but as Elkington (1998) points out, there is little transparency to how this is implemented nor is there adequate agreement on how to audit sustainability. If corporate decision makers are evaluated by their boards of directors on the sole indicator of financial performance on a quarterly basis, thus providing maximum return to stockholders, there is no way that a greener agenda will contribute to this goal unless there are immediate financial rewards.

On the other hand, we must remember that one result of consolidation and large scale is an efficient, industrial and global food industry that supplies us with inexpensive food, a great selection through the entire year, and a certain security based on transportation that moves food long distances from place of production to many markets. With current fuel prices, uninterrupted trade, and high demand for a variety of food products from around the globe, the system works well if we do not assess long-term environmental impacts nor the social inequities that are connected with large disparities in income and access to these products, both within and among countries.

Key Issues in Greening Agriculture and the Food System

When we review these major trends affecting agriculture and food systems in the USA, two key issues arise regarding the greening of agriculture. First is whether farmers who subscribe to the goals of sustainable agriculture strategy and who operate smaller-scale farms do in fact implement practices that are greener than those used in industrial agriculture? The proponents of sustainable agriculture take this as given, while others say that industrial farms can better afford to meet environmental standards. Parallel to this, we explore the impact of business–is smaller really greener? Second, we examine how pursuit of a triple bottom line–economics, environmental agenda, social decisions–impacts the greening of agriculture and the food system. Other issues, such as relative food security with global versus local systems, and what role consumer education has on buying habits and greening of the food system, will wait for another day.

Is Small-Scale, Sustainable Agriculture Greener Than Industrial Agriculture?

To answer this question, we describe Extension experience with those farmers and families that declare adherence to sustainable agriculture. Sustainable farms maintain diversity, individuality, decentralization and local control, and strive for ecological and social integrity and stability. In a family proprietorship or corporation, the decisions reflect the unique social values of that family, as compared with a corporate farm owned by shareholders whose principal or collective goal is return on investment. If capital can provide greater monetary return in another place, the money invested in an industrial farm will leave this farm or leave farming all together to go elsewhere for more lucrative profits (Ikerd, 2005). We define an industrial farm, whatever the size, as one where the operator says "Farming is now a business, just like any other, and the only way to stay in business is to maximize profits."

Proponents of sustainable agriculture often assume that organic farming is a sustainable strategy. Experimental data just published from the farming systems project at the Sustainable Agricultural Systems Laboratory, USDA/ARS, Beltsville, Maryland, compare soil quality in organic and conventional cropping systems (Green et al., 2005). This information comes from the corn year of the crop rotation cycle with corn-soybean. Table 1 contains results of two systems with conventional inputs, no-till (zero tillage) and chisel-till (four tillage opera-

TABLE 1. Soil Quality Characteristics Following Corn in a Two-Year Rotation and Three Farming Systems,[1] USDA/ARS, Beltville, Maryland (Adapted from Green et al., 2005).

Cropping system	Total C g kg^{-1}	Total N g kg^{-1}	Total P mg kg^{-1}	Bulk density g Mg m^{-3}	Aggregate stability mm	Macro-aggregates g g^{-1}
No till	23.6 a	1.8 a	769 a	1.37 b	0.74 a	0.62 a
Chisel till	16.7 b	1.5 b	703 a	1.53 a	0.40 b	0.35 b
Organic	16.5 b	1.5 b	690 a	1.53 a	0.40 b	0.31 b

[1]Numbers in each column followed by the same letter do not differ significantly, $p < 0.05$.

tions), compared with an organic (five tillage operations) system over eight years of the rotation. There is significantly higher total carbon and total nitrogen in the no-till system, and the same level of phosphorus in the three systems. Bulk density is lower while aggregate stability and macro-aggregates are higher in the no-till treatment than either of the other two; these soil-quality factors indicate that no-till is a more desirable system from a soil conservation and quality point of view. According to the authors, tillage seems to be far more important in improving soil physical properties than the organic soil amendments in the organic treatment. Of course, this study of soil properties does not take into account the chemicals applied and their potential for loss in the two conventional systems compared with the organic system. But it appears that no-till management is *greener*.

Soil and nutrient losses from the same study and three treatments over the period of the experiment are shown in Table 2 (Green et al., 2005), with losses predicted by the WEPP model (Flanagan and Nearing, 1995). The no-till and organic treatments had less runoff than the chisel-till treatment, and the no-till treatment had far less predicted soil erosion than either of the other two. The organic treatment had less erosion than chisel-till due to use of more cover crops. The micro-sediment loss was the same for all three treatments. Calculated losses of carbon, nitrogen, and phosphorus were significantly highest in the chisel-till treatment, and lowest in the no-till treatment. The results suggest that a farm that uses no-till practices, chemical herbicides, and synthetic fertilizers can achieve better soil quality even across the larger area that can be farmed, compared with the more labor-intensive, smaller farm using organic methods, including cover crops, animal manure or compost, and multiple tillage operations to manage weeds. The implication is that chemical-intensive farms could contribute to greater soil

TABLE 2. Soil and Nutrient Losses from Fields Under Three Farming Strategies,[1] Predicted by WEPP Model from USDA/ARS, Beltsville, Maryland (Adapted from Green et al., 2005).

Cropping system	Runoff mm yr^{-1}	Predicted soil erosion Mg ha^{-1} yr^{-1}	Micro-sediment %	C kg ha^{-1} yr^{-1}	N kg ha^{-1} yr^{-1}	P kg ha^{-1} yr^{-1}
No till	107 a	85 a	64 a	170 c	15 c	6 c
Chisel till	116 a	64 b	65 a	1074 a	94 a	43 a
Organic	108 a	43 c	65 a	679 b	62 b	29 b

[1]Numbers in each column followed by the same letter do not differ significantly, $p < 0.05$.

sustainability, thus might be *greener* in this one aspect than the farms we classify as having sustainability as a major goal. One of the primary research goals in organic farming is to develop a system where the rotation, competitive ability of component crops, and judicious use of minimum tillage can provide adequate weed management that would contribute to soil quality equal or better than a conventional, no-till farm.

What about succession in ownership of family farms? Sustainable farms often are multigenerational, with land improved and ownership passed on to the next generation within the family. Conventional wisdom is that these owners are more motivated to make sound, long-term decisions. Yet data from a recent Iowa State University study showed that only 27% of all farm operators among the respondents in Iowa had identified a successor, and large farm operators rated family succession higher (3.93 on 5-point scale) than small farm operators (3.46 on the same scale), although the difference was not significant (Nanhou, 2001).

Are smaller farms and businesses greener than larger operations? Again we must examine what motivates farm decisions, and how profit and growth fit into the overall equation. A graduate student with Mike Duffy conducted an extensive survey of the Iowa Farm Business Association members and utilized a Lichert scale to analyze the results (Nanhou, 2001). Farms were classified into three categories shown in Table 3: large farms/high profit, small farms/high profit, and small farms/low profit. Farmers were asked to rank 16 goals and objectives on a 1 to 5 scale. The goals listed are the top five for each of the three groups. Within a group, the first goals marked with an asterisk (*) do not differ statistically. It is clear that making money is a higher goal for

TABLE 3. Relationship of Farm Size and Profit to Farmer Goals (Adapted from Nanhou, 2001).

Farm size/ Profit group	Large farms/ High profit	Small farms/ High profit	Small farms/ Low profit
Most important farm objectives (top five, in descending order of importance)	Making money* Place to raise family* Spend time with family* Seen as good neighbor High production farm	Seen as good neighbor* Spend time with family* Place to raise family* Making money* Being my own boss*	Seen as good neighbor* Being my own boss* Making money* Working outside Working with nature

* Goals followed by (*) did not differ significantly in ratings within each farm category.

operators with large/high profit farms than those with smaller farmers, with either high or low profits. Indeed, when goals are compared across the groups, making high profits is the only goal that is substantially different in rank between the large/high profit farms and the small/high profit farms. We are assuming that smaller farms and sustainable thinking are correlated. When rating production level, the farmers with large operations ranked "high production farm" as number 6.5 of the 16 goals, while the small farm operators ranked this as 11th of the 16 goals. Obviously, small farm operators are less concerned about maximizing production or making more money through higher yields.

In this same study, a pair-wise comparison was made among the 16 objectives to determine if there was a significant statistical difference between the farm groups. In a comparison of the large and small farms, both with high profits, the farmers with large operations rated making money significantly higher (5% level) than the farmers with smaller farms. Other items, such as being viewed as a good neighbor, having a place to raise the family, spending time with the family, being one's own boss, and having a high production farm, were not significantly different between the two groups (Nanhou, 2001).

When asked in this study about whether "being viewed as a conservationist" was important, the average value of the ratings was virtually equal between large and small farmers (Nanhou, 2001). However, while only 5% of large farmers called this issue "very important" a full 45% of small farmers, either high or low profit, called this same issue "very important." Just saying something is very important does not necessarily translate into *greener* practices, but it does indicate a high level of concern.

Are managers of larger farms more likely to comply with environmental regulations than those on smaller farms? Evidence comes from

Wisconsin that operators with farms with over 100 corn acres are more likely than operators with smaller farms to comply with the 20% refuge requirements for planting non-Bt hybrids when using Bt hybrids on most of their ground (Goldberger et al., 2005). In Minnesota, the farms over 200 acres were the most likely to comply with the same regulation. Another dimension of the same study evaluated awareness of integrated resource management as a function of farm size; farms with less than 100 corn acres had an average of 75% awareness, while those with over 100 corn acres had average of 90% awareness among managers who responded to the survey (Goldberger et al., 2005). It appears from this study that managers on larger farms were more concerned than those on smaller farms about environmental issues, or at least were aware enough of the issues to be able to respond that they were taking steps to help the environment.

So are large farms *greener* than small farms? We probably all have anecdotal evidence on both sides of this question and it is difficult to find data to support either conclusion. The paradox of conservation tillage, as in the no-till results from Beltsville just described, is a useful example. Most farmers we meet in the Midwest adopt this practice to meet compliance rules for federal programs, to reduce fuel costs, and to save moisture, especially in the Great Plains. There may or may not be concern about fertilizer or pesticide runoff, and even less about how many neighbors' farms are bought out for the expansion of the larger farms. We observe through our Extension contacts that farmers who adopt sustainable practices, often with smaller farms, have a greater concern for the environment and community, but this is not necessarily because their farms are small–it is their philosophy about management. The conclusions are in agreement with DeVore (2000).

If the singular focus of farming is on profits, industrial-model farms ignore to the extent possible the environmental and social costs of farming. Just as with managers of small farms, the industrial-model managers will comply with those regulations that allow them to reap the benefits of farm support programs. To the extent possible, environmental and social costs will be externalized in order to increase profits. Exporting these costs to society at large creates what Garrett Hardin calls the *tragedy of the commons* (Hardin and Baden, 1977). Likewise, neo-classical economists discount the future, with faith that the market will sort things out. At the same time, sustainable farmers consider more than only short-term profits. Many of the consequences of unsustainable practices by both groups are borne involuntarily by society at large, since they are external to the immediate farming system and there is no formal trading for such "commodities" as ecosystem functions or health effects of chemicals (Tegtmeier and Duffy, 2004). There are methods being developed to quantify the impacts of these factors indirectly through an economic process called ecological contingent valuation, and there may soon be economic ways to reward ecosystem services (Daily, 1997; Costanza et al., 1997) and they will become important in the determination of agricultural and environmental policy.

In response to a question from Mike Duffy on manure management systems and compliance relative to farm size, Iowa Department of Natural Resources Division Administrator Wayne Gieselman said, "With small operations we probably have just as many fish kills as we have from large ones (confined animal feeding operations, or CAFOs). Small operations may cause more 'chronic problems' such as cattle in the stream, and they are sometimes more difficult to deal with because they tend to think that because they are small, they can't be doing much wrong." In contrast, "the big operations can cause catastrophic problems when something does go wrong, but for the most part they are pretty well run and managed. Big operations also are a little easier to deal with simply because they have been in the public eye so much" (e-mail response, 2 November 2005). Mike Duffy concluded after the interview, "It comes down to the question of whether we are willing to accept a low probability of a high-cost problem, or a high probability of many low-cost problems." In the overall picture, we question the dangers of "many small problems." Potential pollution can result from manure concentrated in one area, for example, yet manure becomes a valuable resource for maintaining soil fertility when that by-product is dispersed. One of the obvious problems with the industrial model and larger operations is the concentration itself. Large numbers of animals in a small area lead to concentration of manure and difficulties of disposal, compared with spatial dispersion of the same by-product that then becomes a resource. There are also animal health problems that do not occur with dispersed, grazing livestock.

In contrast to the information in this email from an administrator in Iowa, a study of eastern and western Cornbelt hog farms that includes 86% of the farms with confined hogs in the USA highlights the difference between potential and practice (Ribaudo et al., 2003). In their results, 85% of the confinement units in the eastern Cornbelt with less than 300 animal units had adequate land for applying nitrogen from hog manure, while only 56% with more than 1000 animal units had adequate land. In the western Cornbelt, 92% of the smaller units had adequate land while 67% of the larger units had enough land to spread manure within the federal guidelines. The results suggest that smaller confined animal operations have greater potential to meet environmental compliance than larger units. The study further reported, "While many small- and medium-size farms control enough land to meet nutrient standards, most are not applying manure on all of their cropland, thereby over-applying manure on the acres they do use." The authors concluded, "As the size of animal operations increases, nutrient imbalances also typically increase . . . mainly due to a lack of proper land area for spreading manure." Thus there is greater potential for small farms to be *greener*, but we cannot tell from this study that in fact they have a higher level of compliance. Moreover, the large concentrated animal feeding units have a huge impact on the industry and potentially on the environment, since they are only 5% of the feeding operations but include 50% of all animals and produce 65% of the excess nutrients. Thus it is difficult to reach clear conclusions on the impact of farm size on the environment. We know from experience that small and large farms have greater application rates near the animal facility, since this is more convenient and time is often a constraint. Efficient use of time for spreading can easily result in over-application and potential for environmental pollution, on both small and large farms.

Similar to locally own and farmer-managed farms, local businesses owned by people in the community are likely to be *greener* and to stay in business, especially if local residents support those entrepreneurs. When ownership, management, and the work force are all local, there is commitment to the community, and more than a single bottom line. There is less incentive for the company to move elsewhere in the USA or offshore to seek lower labor costs than if the company is owned by stockholders who only look at the quarterly dividend or price of stocks. In a local business, when we add the value of labor to the raw materials or products that are sold, most of this value stays in the community. The farmer buys fertilizer; the fertilizer dealer pays local people to help, and they buy shoes. The owner of the shoe shop buys some local vegetables from the food market, and hopefully some of these come from the local farm. Local food has an identity and a face. When there is more than a single motive of profit for the business, that family and business will stay in the community. Local does not guarantee *green-ness*, but it does promote stability and security in the system at the local level.

Wal-Mart is currently the largest food retailer in the USA (Heffernan and Hendrickson, 2005). A few years ago a visiting farmer from Texas described how Wal-Mart destroyed business in their rural community of 8,000 people not once, but twice. When the big box retailer came into the community, it created a flush of jobs for minimum wage labor, although the contractor was from a distant city and most of the building materials were imported. The store created a number of low-wage positions for sales people; the scale, efficiency, and lower prices forced the majority of small, local shops out of business. Thus the super store replaced moderately paid entrepreneurs with minimum wage clerks and stockers. Five years later, a business decision from a distant management team deemed the large store unprofitable and it closed without prior notice. Even the low wage jobs were lost, and it was extremely difficult for any of the small shops to reopen as buying habits of consumers had changed and people now drove to super stores in other towns. Again, local ownership does not guarantee *green-ness*, but it keeps more money in the local economy and local people are accountable to their neighbors.

No discussion of impact of farm and business size on community is complete without citing the landmark research of Walter Goldschmidt in the southern San Joaquin Valley of California in the late 1940s (Goldschmidt, 1947). He compared two rural towns about the same size, Dinuba and Arvin, with similar soils, irrigation systems, and production potential. For historical reasons, Dinuba was a town surrounded by small family farms with average size of 57 acres dedicated to dairying and other livestock, growing fruits and nuts, with diverse field and forage crops and mixed farming systems. A few miles away, the town of Arvin was surrounded by large, industrial-model farms with average size of 497 acres with cotton and other extensive crops. Annual incomes were $3,300 on small farms and $18,000 on large farms. In his study of local business and organizations, Goldschmidt found a striking difference. In comparison with Arvin, Dinuba had twice as many self-employed people, five schools rather than one, three public parks compared with one small playground, two newspapers compared with one, and twice as many civic organizations and church groups. Crime rates were lower in Dinuba, and there was an obvious difference in quality of life, whatever the indicators used. Goldschmidt concluded that small farms, locally owned and operated, were the foundation for a healthy, stable, and safe community compared with the industrial alternative that employed migrant labor and lacked the same stability and breadth of ownership. California Farm Bureau Federation hotly opposed his results at the time, and that group representing large farms tried to suppress the results and publication, but the research was completed and reported. The findings of Goldschmidt were confirmed by Dean McCannell (1991) in a number of communities in several states.

The greater local investments, small business successes in farming and the community, and security and stability of the food production system all undoubtedly contributed to a greener community, although demonstrating this was not a goal of the study.

In spite of the lack of clear evidence that relates *greening* to size of business, there are tools available to make such comparisons. Life cycle analysis of products is a method that includes an accounting of all energy and material flows through the life of a product, from design and manufacture through sale, use, and disposal. The calculations are complicated and depend on many assumptions when data are scarce. A description of the process from the Environmental Literacy Council is available (http://www.enviroliteracy.org/article.php/322.html; accessed 18 November 2005). Local food sold in the community has far less transportation costs than food grown and packaged and sent 1500 miles (average food distance in the U.S.) to the consumer, although efficiencies of scale in processing and packaging may cancel the travel costs in the short term with current energy prices. There may be obvious waste in a farmers' market, for example unsold products, but some of these go to food banks and spoiled products often return to the farm for feeding livestock or composting. Waste in the larger industrial food system is less visible, and costs may be externalized to society. A small business, be it the cooperative selling fertilizers and pesticides or the local food market, has an immediate local accountability because most of its activities are visible and known to the community. If there is an environmental problem, the responsibility is with someone we know and not a distant corporate management entity somewhere else. In a movie, *The Grapes of Wrath*, the man in the large convertible coming to foreclose on the Joad family and neighbors said he was just following orders, his boss was accountable to the bank, and the local bank to some stockholders off in another city. Essentially no one was responsible, when the poor irate farmer with a shotgun said, "Well then, who CAN I shoot?" (adapted from the novel by Steinbeck, 1939). We conclude that the smaller, local and responsible ownership has higher probability of *greening* the community.

Do Economics, Environmental Issues, and Social Decisions Impact Greening in Agriculture?

We approach this question using data from the organic food production and marketing sector in the USA. Although still a small fraction of the total U.S. food supply, organic foods have been growing at about

20% per year for the past two decades. Worldwide, there are currently 26 million certified organic hectares, or 66 million acres (Willer and Yussefi, 2005). There is strong interest in the industrial farm and mass marketing business in taking advantage of this growing sector of the food business, since, in general, food purchases are relatively inelastic with changes in income, at least in the developed countries. Figure 3 illustrates the number of organic farms in the USA in 2002 by size category, as well as their percent contribution to the total amount of organic food produced. The figure shows that about 200 farms, 1.5% of the total, produce 45% of organic products (USDA, 2004).

Given the strict rules that organic farms must follow, including prohibition of application of chemical pesticides and fertilizers, we assume that most organic farms are *greener* than most conventional farms, although we recognize that more tillage for weed management may contribute to soil erosion. When most organic farms were small and provided vegetables, fruit, meat and other products for primarily a local market, and before there was a national organic certification standard, this was a segment of the food sector that had strong philosophy about the environment and social motivation. Most labor was from the family, and benefits accrued to those who were closely involved. Today, in the opinions of many of the founders of organic agriculture, what was once a unique segment of the food industry is moving into the industrial mode and beginning to resemble large-scale commercial farming and

FIGURE 3. Value of Sales and Number of Farms by Farm Size in 11 Categories in the USA in 2002 (USDA, 2004).

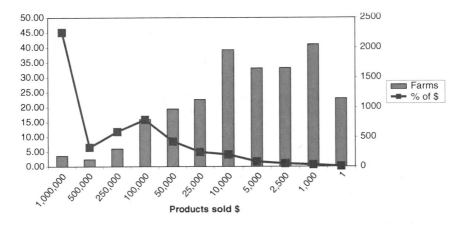

food systems in most ways. Some of the original organic farmers are looking for new ways to certify local foods grown by people with strong social motivation that will reach a wide range of consumers, not just economic elite.

Most people believe that the environment should be a concern for us all. Some people and organizations deal with this concern through in-depth research and changes in activities, a true *greening* of their behavior including decisions in farming and the food system. Others *greenwash* their activities by putting an environmentally friendly face on a company or a product in order to differentiate theirs from others, and to take advantage of general society concern. *Greening* of agriculture is far more than a simple study of the marketplace and expanding the advertising process to wrap a product in environmental terms. *Greenwashing* is to express concern about the environment, saying in advertising that the environment is given special consideration, or is a primary motivation, when in fact it is not. There is no question that individual economic and social decisions can impact the greening of a country and its economy. The major challenge for the consumer is to sort out the truth of a situation for a given product and separate that from the advertising, and especially the *greenwashing* of the image of a product.

CONCLUSIONS

Based on the evidence from the literature, discussions with farmers and business owners, and long-term Extension experience of the four authors, we draw several conclusions about the questions posed in the introduction. As with any complex issue, there are conflicting data and ideas about whether farm or business size and ownership pattern are related to the *green-ness* of agriculture and the food system. It is self-evident that if we pursue a single bottom line, such as economic profit, we are likely to have more success achieving that goal than if there are multiple objectives, such as a triple bottom line. Here are our conclusions.

Does Scale of Farming or Business Relate to Green-Ness and Sustainability?

We conclude that the sustainable strategy for agriculture, both in production and the rest of the food chain, is inherently *greener* as we define the term because of the triple bottom line of profitability, environmental soundness, and social responsibility. The sustainable farming strategy includes a concern for environmental soundness that is especially expressed in the immediate landscape, and also as a part of the farmer's social responsibility to the local community. A sustainable farm must also be profitable in current economic terms, but that is not the only goal. We have shown above that small farmers are somewhat less concerned about making money and about high yields. Yet three of four farmers have no identified successor, similar across farm size, so the multi-generation issue is inconclusive. Soil management and erosion control are better with no-till than with either organic or conventional-till practices, although the cover crops in organic systems reduce soil erosion from the levels found in conventional practices.

As a general conclusion, we observe that many characteristics of smaller family farms and local businesses are *greener* as we define the term than industrial farms and multinational corporations. To be sure, there are some small farms and local businesses that are not at all sustainable, and there are some large operations that could be called sustainable if they are serious in pursuing a multidimensional bottom line. We find some conflicting data. For example, small hog operations have better manure management potential but perhaps worse practices than larger operations; small-acreage corn farmers have lower compliance rates for Bt-refuge requirements than larger operations. If any business subscribes to a multiple bottom line, regardless of size, it is more likely to be *greener* than one that does not.

Do Economics, Environmental Issues, and Social Decisions Impact Greening in Agriculture?

We conclude that economic, environmental, and social issues all influence how *green* our agriculture and food systems can be. Some of the environmental implications of farming decisions with relation to choice of tillage practices, use of cover crops, timing and rates of application of nutrients, and judicious use of pesticides can affect how *green* a farming operation will be, regardless of size. Economic incentives and regulations on farming practices also impact decisions, and whether farmers follow these rules will make the operation more or less green. In business the sourcing of products, including concern about how and where they were produced, the handling of waste, and the use of energy, all can contribute to a more or less green enterprise. Again, it is certain that a farm or business manager who is concerned about environmental soundness and is aware of the social impacts of that business activity, in addition to concern about economics, will be *greener* than a farm or business that is focused only on the economic bottom line.

We have presented the evidence easily available on both small and large farms, sustainable and industrial operations, and managers and organizations that are pursuing single and multiple bottom lines. There are undoubtedly many more information resources that could bias the conclusions in one way or the other about the *greening* of agriculture and the food system. We hope that this discussion provides insight into the issues, and helps the informed reader ask the right questions and sort out many of the conflicting claims in advertising. Everyone agrees on the need for a *greener* future, and we mainly differ on strategies to achieve that desirable future situation.

REFERENCES

Allen, P. 2004. Together at the table: Sustainability and sustenance in the American agrifood system. Pennsylvania State Univ. Press, Univ. Park, Pennsylvania.

Cochrane, W.W. 2003. Curse of American agricultural abundance: A sustainable solution. Univ. Nebraska Press, Lincoln, Nebraska.

Costanza, R., C. Perrings, and C.J. Cleveland, editors. 1997. Development of ecological economics. Edward Elgar Publishing, Camberley, Surrey, UK.

Daily, G. (editor). 1997. Nature's services: Societal dependence on natural ecosystems. Island Press, Washington, DC.

DeVore, B. 2000. Bigger is better and three other agricultural myths. In Motion Magazine, February 6.

Duffy, M., and D. Smith. 2004. Farmland ownership and tenure in Iowa 1982-2002: A twenty year perspective. Iowa State Univ. Extension, PM 1983, July, Ames, Iowa.

Elkington, J. 1998. Cannibals with forks: The triple bottom line of 21st century business. New Society Publishers, Stony Creek, Connecticutt, USA.

Flanagan, D.C., and M.A. Nearing, editors. 1995. USDA Water erosion prediction project: Hillslope profile and watershed model documentation. NSERL Report #10, USDA-ARS National Soil Erosion Laboratory, West Lafayette, Indiana.

Francis, C. 2004. Greening of agriculture for long-term sustainability. Agron. J. 96:1211-1215.

Francis, C.A., C.B. Flora, and L.D. King, editors. 1990. Sustainable agriculture in temperate zones. John Wiley & Sons, New York.

Goldberger, J., J. Merrill, and T. Hurley. 2005. Bt corn farmer compliance with insect resistance management requirements in Minnesota and Wisconsin. AgBioForum, 8(2&3):151-160.

Goldschmidt, W.R. 1947. As you sow. Harcourt, Brace and Co., New York.

Green, V.S., M.A. Cavigelli, T.H. Dao, and D.C. Flanagan. 2005. Soil physical properties and aggregate-associated C, N, and P distributions in organic and conventional
cropping systems. Soil Science 170(10):822-831.

Hardin, G., and J. Baden. 1977. Managing the commons. W.H. Freeman, San Fran-
cisco, California.

Heffernan, W., and M. Hendrickson. 2002. Concentration in agricultural markets. National Farmers Union, http://www.agribusinessaccountability.org/page/149/1.

Heffernan, W., and M. Hendrickson. 2005. The global food system: A research agenda.
Conf. Corporate Power in the Global Food Systems, High Leigh Conf. Centre,
Herfordshire, UK. 19 p.

Ikerd, J. 2005. Sustainable capitalism: A matter of common sense. Kumarian Press,
Bloomfield, Connecticut.

Ikerd, J. 2006. Contradictions of principles in organic farming, In Organic agriculture: A global perspective, P. Kristiansen and A. Taji, editors. CSIRO Publishing,
Collingwood, Australia, pp. 221-229.

Levin, R.A. 2003. Willard Cochrane and the American family farm. Univ. Nebraska
Press, Lincoln, Nebraska.

Lieblein, G., E. Østergaard, and C. Francis. 2005. Becoming an agroecologist through
action education. J. Agricultural Sustainability (U.K.) 2(3):147-153.

MacCannell, D. 1988. Industrial agriculture and rural community degradation.
pp. 15-75, 325-355. In L.E. Swanson (ed.) Agriculture and Community Change in
the U.S. Westview Press, Boulder, CO.

Morgan, D. 1979. Merchants of grain. Viking Press, New York.

Nanhou, V.Y. 2001. Factors of success of small farms and the relationship between financial success and perceived success. M.S. Thesis, Department of Economics,
Iowa State Univ., Ames, Iowa (unpublished).

Pizzigati, S. 2004. Greed and Good: Understanding and overcoming the inequality that
limits our lives. The Apex Press, New York.

Ramos, C.A., and I. Monticl. 2005. When are corporated environmental policies a form
of "greenwashing"? Business & Society 44(4):377-414.

Ribaudo, M., N. Gollehon, M. Aillery, J. Kaplan, R. Johansson, J. Agapoff, L.
Christensen, V. Breneman, and M. Peters. 2003. Manure management for water
quality. Costs to animal feeding operations of applying manure nutrients to land.
Agricultural Economic Report 824, Economic Research Service, U.S. Dept. Agriculture, Washington, DC.

Schlosser, E. 2001. Fast food nation: The dark side of the all-American meal. Houghton Mifflin, New York.

Steinbeck, J. 1939. The grapes of wrath. Viking Press, New York.

Tegtmeier, E.M., and M.D. Duffy. 2004. External costs of agricultural production in
the United States. Intl. J. Agric. Sustainability 2(1):1-20.

Troyer, A.F. 2004. Background of U.S. hybrid corn. II. Breeding, climate, and food.
Crop Science 44(2):370-380.

U.S.D.A. 2004. Agricultural statistics, 2004, U.S. Dept. of Agriculture, Washington,
DC.

U.S. Department of State. 2002. Food security and safety. Economic Perspectives,
Washington, DC. Vol. 7, No. 2, May.

Waage, S., and C. Francis 2004. The elephant in the room: Core stumbling blocks to operationalize sustainability. Green Money Journal, Summer. Available at: http://www.greenmoneyjournal.com/article.mpl?newsletterid=29&articleid=317

Willer, H., and M. Uussefi, editors. 2005. The world of organic agriculture: Statistics and emerging trends, 2005. IFOAM, Bonn, Germany.

doi:10.1300/J411v19n01_10

Index

Note: Page numbers in *italics* refer to figures or tables.

BOOK ORDER FORM!

Order a copy of this book with this form or online at:
http://www.HaworthPress.com/store/product.asp?sku= 6023

Agricultural and Environmental Sustainability
Considerations for the Future

___ in softbound at $65.00 ISBN-13: 978-1-56022-171-5 / ISBN-10: 1-56022-171-2.
___ in hardbound at $80.00 ISBN-13: 978-1-56022-170-8 / ISBN-10: 1-56022-170-4.

COST OF BOOKS _____

POSTAGE & HANDLING _____
US: $4.00 for first book & $1.50
for each additional book
Outside US: $5.00 for first book
& $2.00 for each additional book.

SUBTOTAL _____

In Canada: add 6% GST. _____

STATE TAX _____
CA, IL, IN, MN, NJ, NY, OH, PA & SD residents
please add appropriate local sales tax.

FINAL TOTAL _____
If paying in Canadian funds, convert
using the current exchange rate,
UNESCO coupons welcome.

❑ **BILL ME LATER:**
Bill-me option is good on US/Canada/
Mexico orders only; not good to jobbers,
wholesalers, or subscription agencies.

❑ **Signature** _____

❑ **Payment Enclosed: $**_____

❑ **PLEASE CHARGE TO MY CREDIT CARD:**

❑ Visa ❑ MasterCard ❑ AmEx ❑ Discover
❑ Diner's Club ❑ Eurocard ❑ JCB

Account #_____

Exp Date_____

Signature_____
(Prices in US dollars and subject to change without notice.)

PLEASE PRINT ALL INFORMATION OR ATTACH YOUR BUSINESS CARD

Name

Address

City   State/Province   Zip/Postal Code

Country

Tel   Fax

E-Mail

May we use your e-mail address for confirmations and other types of information? ❑Yes ❑No We appreciate receiving
your e-mail address. Haworth would like to e-mail special discount offers to you, as a preferred customer.
We will never share, rent, or exchange your e-mail address. We regard such actions as an invasion of your privacy.

Order from your **local bookstore** or directly from
The Haworth Press, Inc. 10 Alice Street, Binghamton, New York 13904-1580 • USA
Call our toll-free number (1-800-429-6784) / Outside US/Canada: (607) 722-5857
Fax: 1-800-895-0582 / Outside US/Canada: (607) 771-0012
E-mail your order to us: orders@HaworthPress.com

For orders outside US and Canada, you may wish to order through your local
sales representative, distributor, or bookseller.
For information, see http://HaworthPress.com/distributors

(Discounts are available for individual orders in US and Canada only, not booksellers/distributors.)

Please photocopy this form for your personal use.
www.HaworthPress.com

BOF07